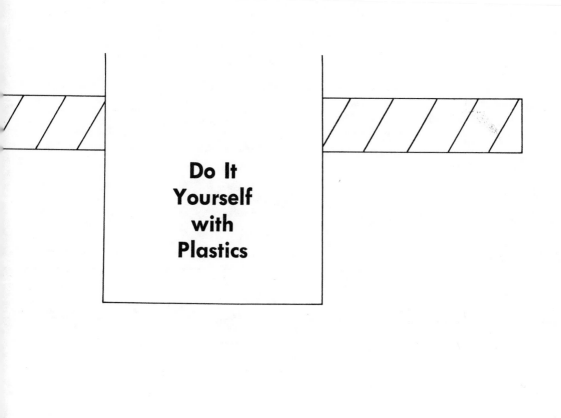

Do It
Yourself
with
Plastics

A Sunrise Book E. P. DUTTON & CO., INC. · NEW YORK

ERICH H. HEIMANN

Translated by the author

Do It Yourself with
PLASTICS

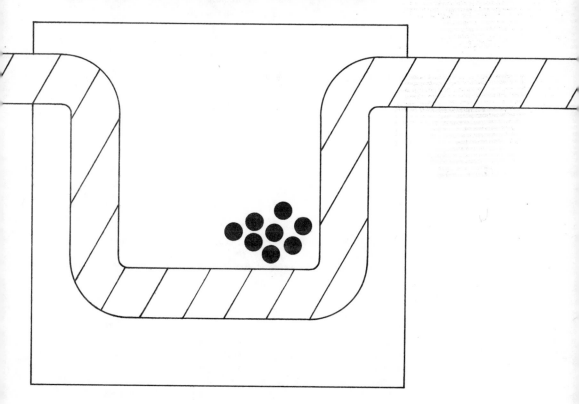

DUTTON-SUNRISE, INC.,
a subsidiary of
E. P. DUTTON & CO., INC.

Library of Congress Cataloging in Publication Data

Heimann, Erich Hermann.
Do it yourself with plastics.

Translation of Aus Kunststoff selbst gemacht.
"A Sunrise book."
Includes index.
1. Plastics. I. Title.

TA455.P5H39513   1975   668.4'9   75-12915

ISBN: 0-87690-178-X

Designed by The Etheredges

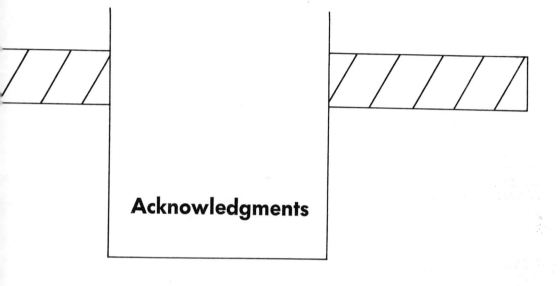

Acknowledgments

The publishers and the author warmly thank the following persons and companies for their kind assistance in supplying technical information and illustrations: Arbeitsgemeinschaft Deutsche Kunststoffindustrie (AKI),* Frankfort; Badische Anilin- und Soda-Fabrik AG (BASF), Ludwigshafen; Bayer AG, Leverkusen; Bostik GmbH, Oberursel; Dynamit-Nobel Aktiengesellschaft, Troisdorf; Emerson & Cuming Inc., Canton (Mass.); Farbwerke Hoechst AG, Frankfort-Hoechst; Haellmigk-Kunststoffe, Augsburg; Konrad Hornschuch AG, Weissbach; Kalle AG, Wiesbaden; Mr. G. Neuhaus, Frankfort (Main); Mr. K. B. Schoenenberger, Kattenhorn; Omni-technik GmbH., Munich; Teroson GmbH., Heidelberg; Vosschemie, Uetersen/Bondaglass-Voss Ltd., Beckenham (Kent).

Thanks are also due to Miss Jo Hertz and Mr. Peter Owen for reading the English script and making many helpful suggestions.

* Society of German Plastics Industry.

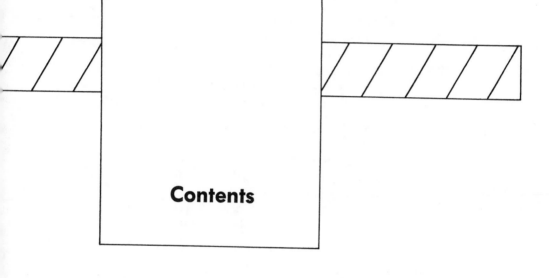

Contents

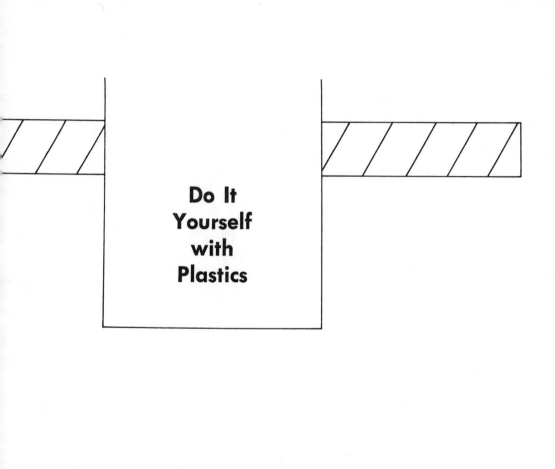

**Do It
Yourself
with
Plastics**

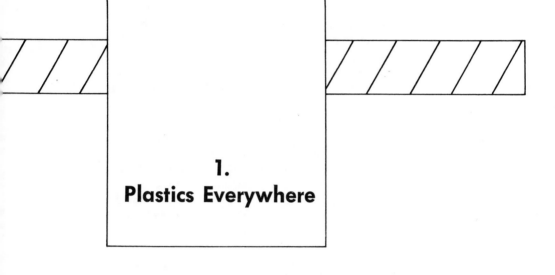

1.
Plastics Everywhere

Probably no other species of materials has gained such a quick and wide application in the technical world and in everyday life as plastics have done; yet, at the same time, most of its users know little of their nature and properties. Today, it is hard to find a household in which plastic products are not in everyday use. Bowls and tableware, the casings of vacuum cleaners, handles of knives, pots and pans, furniture covers, tabletops, the insulation of the refrigerator, bags and containers in deep-freeze and picnic baskets, the non-stick coating of saucepans and ovenware, and many other things that come from the retorts of the plastics industry. A modern car contains about 90 lb. (40 kg.) of different plastics. Plastics serve as electrical insulations, such as coverings of wires and cables, or as battery cases, making them safe and comfortable to touch. Plastics offer easily cleaned interiors; they are used to pad dashboards and the hubs of steering wheels to reduce risks of injury in accidents. Recently developed bumpers molded from resilient plastic help to prevent injuries to passengers and damage to the car in low-speed collisions or parking mishaps. Injected into the cavities of car bodies and chassis, plastics banish internal corrosion and, at the same time, reinforce and soundproof the body work.

In modern aircraft, about 40 percent of structural joints are bonded with plastic resins, resulting in stronger and better joints than bolting or riveting. Even the Jumbo Jet might never have flown if light yet strong plastics had not reduced the weight of the "flying elephant" by about ten tons. Plastics have proved indispensable in the exploration of space and in many other important fields.

But much nearer home, plastics have conquered in a fairly short time. They have even managed to enter the field of fashion, at first, in the shape of buttons and clasps, and not long afterwards, as films (or foils) and plastic-coated fabrics. Bright colors and glossy surfaces banished the dull gray "mac" and created rainwear fashions. Even the French fashion houses, led by Courreges, discovered the new material and used it for their creations. Films (or foils) and plastic-coated fabric joined the synthetic fibers, which had already gained the favor of the textile industry and of the ladies, too, who learned to appreciate the advantageous properties of synthetics. This trend led to the development of new easy-care materials combining high strength and attractive finish with another very important quality: they are watertight but do not exclude air. Coated fabrics with a glossy textured finish and other leatherlike materials set new trends in fashion and in many fields of application even succeeded in partially ousting textiles and leather. Of low weight and high strength, these new materials are widely used for suitcases and bags. As covering materials for upholstered furniture they excel in their resistance to stains and in their quick and easy maintenance.

In the field of sports and hobbies, we also meet plastics in great variety. Motor boats and garden furniture, swimming pools and precision-modeled model railways, the running track in a modern stadium, the flexible poles of record-breaking pole vaulters, who now vault to heights never dreamed of before—all these items are made from plastics. In the past five years, the world production of plastics has nearly doubled. In 1970, the consumption of plastics per capita reached a record, with approximately 119 lb. (54 kg.) in West Germany. The Americans were runners-up, with 92 lb. (42 kg.).

2.
Plastic
Materials
with
Rules of Their
Own

In view of the consumption of plastics, it is astonishing how little is known about them. Some people still think of plastics as cheap, not very durable and more or less ersatz—a substitute—while others are most impressed by highly specialized plastics, which were developed at vast expense for special applications in the field of astronautics. Both points of view lead to a wrong view of synthetic materials, which have progressed far beyond the cheap substitute stage but, on the other hand, cannot be used anywhere and everywhere with equal success, even though some of them excel by astonishing and convincing performances.

Success depends on the proper use of the right type of plastic material in the right place.

This very much looks like a commonplace—a self-evident

rule that should be applied to each and every material—but even in very recent times, this rule was often neglected. This led to failure, which was blamed on the plastics instead of on the way they were used. One company, for instance, copied the design of a tin watering can, which had been assembled by soldering several tin parts together. The plastic version was an exact copy, but was injection-molded in one piece. The result of this not very ingenious performance: after a short time in use, the handle broke because it met the body of the can at a too acute angle, so that when the can was filled with water, the load could not be evenly dispersed over a large enough area, as is required with this material. Faulty designs, which impaired the image of plastics, have fortunately become quite rare, though now and then one still sees products where design and material are at variance. A material specifically designed for one purpose may offer no advantages when put to a different use. The same applies if material is wrongly treated. High-gloss polyester furniture will be spoiled when treated with abrasive cleaning agents, and a polypropylene (e.g., Propathene) bowl turns into a malodorous mess when put on a hot plate. The enormous range of plastics, with its wide variety of special properties, offers materials made to measure for many applications, guaranteeing perfect results if they are used in the right way and place.

It is the purpose of this book to show how to use the different types of plastics in the right way, what can be expected of them, how to repair them, and how they can be used to repair other materials. But even if you are not much interested in technical or practical applications of plastics and prefer artistic and creative activities, you may achieve many new and impressive effects and, thus, find undiscovered aspects of a subject, which, at first, might appear rather prosaic.

DIFFERENT TYPES OF PLASTICS AND THEIR PROPERTIES

We have only to look around to see that plastics can have many different properties and present themselves in just as many different formulations.

Chemical Building Stones

As a rule, they have one thing in common: they contain carbon atoms as basic building bricks, to which some other elements

—principally hydrogen, oxygen and nitrogen, sometimes also chlorine—are linked. A further common trait of synthetic materials is the extraordinary size of their molecules, in which thousands of carbon atoms are linked to each other, as well as to the elements mentioned above. The building of such giant molecules takes place mainly by the joining of small reactive basic units, called monomers. They can also originate from the variation or combination of natural substances, which already are very large molecules, or macromolecules, to call them by their technical name. This happens, for instance, with all plastics produced by transforming the natural product cellulose—a process first tried and perfected in the nineteenth century.

Carbon atoms

The role of the linking carbon atom can also be taken over, in some plastics, by a different kind of atom. In practice, however, only one group of plastics of this kind, the so-called silicones, have gained practical importance. Here silicon atoms play a similar role to that of the carbon atoms in other plastics. In addition, research laboratories in the U.S. and the U.S.S.R. have been experimenting with new plastics in which aluminum atoms act as links in the molecular framework. Due to their very high production costs and their special properties, these materials are often called "exotic" plastics. Together with silicones, they form the basis of a sideline of plastics, which is highly interesting from the technical and chemical point of view: the "inorganic polymers." Excepting the silicones, these plastics have not yet acquired any practical importance.

Silicon atoms

New: aluminum atoms

Properties Depend on Structure

For practical purposes, the enormously varied properties of plastics are much more interesting than the macromolecular structure they have in common.

Three groups of plastics

In essence, we have to distinguish between these basic groups of plastics: first, the so-called *thermoplasts,* which are mainly represented in the shape of films or foils and moldings; second, the *duroplasts,* used for very large moldings, such as motorboats or car bodies, but also for smaller parts, such as handles for pots and pans or cases for electrical switches. They also serve as coating materials for concrete or metal and as special adhesives with very high bonding power. Last, but not least, there are the so-called *elastomers,* which are mainly used as fillers for cracks, mold-making material or padding.

These three main groups may be roughly distinguished by

a rule of thumb: thermoplastics can be molded by the use of moderate heat, while duroplastics, once hardened, cannot be re-shaped by heat. Elastomers are rubberlike elastic materials.

THERMOPLASTIC MATERIALS

Mainly used for cheap, useful articles

The majority of all mass-produced plastics used for making cheap articles for everyday use and industrial products belongs to the thermoplastic group, which are also called "plastomers" in professional terminology. Well-known examples are polyvinyl chloride (PVC), which is available in both a rigid and a flexible form; polystyrene, familiar from a great many smaller household items, such as measuring cups and butter dishes, but also as containers and plastic model kits; and polyethylene or polythene, which we often meet in the shape of plastic shopping bags. Polythene is also used for bottles and other containers.

PVC in its soft and flexible form is offered as plastic film or foil, as artificial leather, as garden hoses, as a lightweight foam, as a self-adhesive decorative film or foil and as a flooring material —just to mention a few fields of application. Rigid PVC is used for plastic pipes and panels and, as a thin film or foil, for lining "paper" and padding material. Other thermoplastic materials

MAXIMUM USABILITY TEMPERATURES

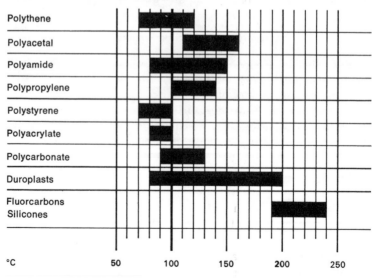

Polythene			
Polyacetal			
Polyamide			
Polypropylene			
Polystyrene			
Polyacrylate			
Polycarbonate			
Duroplasts			
Fluorcarbons Silicones			

°C   50   100   150   200   250

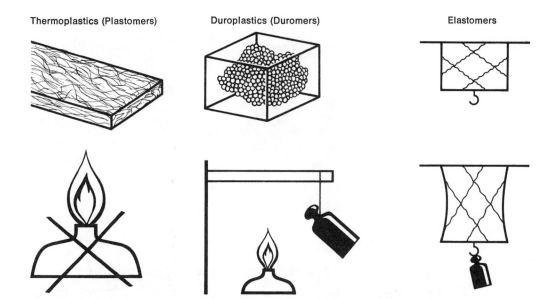

Thermoplastics (Plastomers)   Duroplastics (Duromers)   Elastomers

The molecular structure determines the properties. Thermoplastics have a feltlike structure consisting of long thread-type molecules and are softened by heat (left). Duroplastics form three-dimensional netlike macromolecules and resist heat fairly well (center). Elastomers also form three-dimensional netlike macromolecules with a much wider mesh than that of the rigid duroplastics. This is the reason why elastomers are elastic.

are acrylic resins, the best known of which is Plexiglas (or Perspex), an acrylic glass. Polycarbonates can also be used to make glasslike panes and moldings. A well-known representative of this group is Makrolon, a thermoplastic material that is used for the production of unbreakable windowpanes and for household articles that withstand fairly high temperatures, and even resist boiling water. Some other thermoplastics are used mainly in the technical field: polypropylene (for laboratory equipment and surgical equipment, but also for containers for household use); polyamides, from which fibers as well as technical articles, such as bearings and gear wheels, are made and, finally, polyacetals used for gear wheels, casings (housings) and similar items.

All thermoplastics consist of threadlike macromolecules, the length of which is about one hundred to one thousand times greater than their diameter. These threadlike molecules form a feltlike structure and are kept in place by physical forces, which are the result of mutual attraction between the single molecules. In addition to that, the molecules are held together by a three-

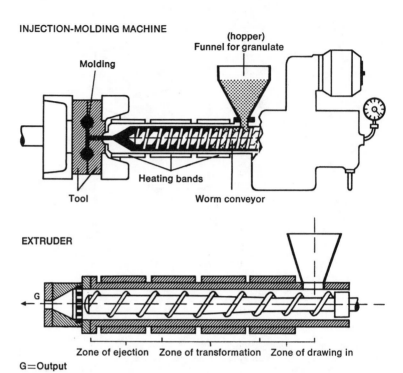

INJECTION-MOLDING MACHINE

Molding

(hopper)
Funnel for granulate

Heating bands

Tool

Worm conveyor

EXTRUDER

G

Zone of ejection   Zone of transformation   Zone of drawing in

G=Output

dimensional interlocking of molecules similar to that of several lengths of tangled string.

Heat makes thermoplasts moldable

This merely physical cohesion can be disturbed by physical means. If the material is exposed to heat, the threadlike molecules begin to oscillate. These oscillations are antagonistic to the forces that keep the molecules together. The molecular structure begins to loosen. The material loses its rigidity and strength, softens and finally melts.

If the melted material is cooled down again, the oscillations of the molecules are progressively reduced and the binding forces prevail, so that the material becomes rigid again. This effect is utilized in technical practice. Heat-softened plastomers can be molded. Once cooled again, they keep their new shape.

Deep drawing

This molding process can also be achieved by deep drawing, when the softened plastic sheet is drawn over a mold by means of a vacuum. The material cools on the mold and can then be removed to trim off any excess material. This process is mainly used in the production of more or less two-dimensional parts (bas-reliefs), such as decorative panels for walls and ceilings.

Another field for deep drawing is the clear blister pack, for tablets or goods sold in supermarkets. Deep drawing requires special machines and equipment for perfect results. If the quality of the molding does not have to be perfect and if the drawing depth is not too great, it is possible to imitate the technical process of deep drawing at home in your own workshop (see pages 210–212 for detailed description).

The two remaining molding processes, injection molding and extrusion, are of interest only to industry. This also applies to the blow-molding process, which is used for the production of films or foils and bottles and containers of all kinds. For injection molding and extrusion, the machine is fed with the plastic raw material in the shape of beads (granulate), which are steadily heated in a metal chamber from which the material is transferred to the steel mold by means of a worm conveyor, which is similar to that of a mincer. Once injected into the completely closed mold chamber (injection-molding) or open-calibrated molding channel (extrusion), the plastic material solidifies on cooling, which is generally accelerated by cooling the mold with water. In injection molding, the mold is then automatically opened, and the molding is ejected. The mold then closes again and is now ready for the next cycle.

Injection molding, extrusion and blowing: industrial processes

Quite often, the mold contains several chambers, so that several identical or different parts can be molded in one operation. The moldings are then joined by "spigots" and ejected in one piece. The single parts can be easily broken off afterwards. As injection molds are high-quality precision tools, which may cost tens of thousands to produce, they are economical only for mass production. Amortized over several hundred thousand moldings, the tool costs are no longer of great significance. The parts leave the machine completely finished, and there is no need for any manual work, except to remove the spigots, so the cost of such parts is generally fairly low. This means that for the handyman, it is generally cheaper to buy a new part than to fiddle around with complicated rejoins.

Rather than repair an injection molding, buy a new one

Unlike injection molding, extrusion is not a cycle process but a continuous process. Here, the softened plastic mass is pressed through a calibrated channel, where it is already cooled to an extent, which ensures that the material has returned to its rigid state by the time it leaves the machine. Mainly profiles and hoses are produced by extruding.

Blow molding is used to make hollow parts (e.g., bottles)

and tubular film or foil for packing. Hollow articles are molded in a hollow coreless mold (female mold). The semimelted (plastic) material is blown up to a balloon with a fairly thick skin. The balloon is formed by compressed air, which also presses the still moldable skin against the walls of the female mold, where it cools and solidifies.

When blowing plastic films or foils, which is a continuous process similar to extruding, the film, which is plasticized by heat, is pressed through a horizontal annular gap. Injected air widens the at-first still rather thick-walled "tube," which is sealed at its upper end, to an inflated balloon with a very thin skin. This balloon stands vertically above the blowing machine and automatically cools as it is exposed to the surrounding air. Thus, the blown tubelike film bag enters its solid state and can be wound up to form a large coil. Unfortunately, this process cannot be copied at home.

How to Treat and Process Thermoplastics

The very property that makes them so easy to mold can be a disadvantage in the use of thermoplastics: they are fairly sensitive to heat. With some thermoplastics 140–58° F (60–70° C) is the maximum they will stand (e.g., some types of polythene and polystyrene). This means that really hot dishwater may be enough to cause warping or to turn clear parts opaque. Thermoplastics take an even dimmer view of being put on a hot plate. An immersion heater must not be put into a plastic container.

The diagram on page 6 shows the maximum temperatures that different types of plastics can withstand. At certain temperatures, thermoplastics are deformed. If the temperature is raised further, an irreversible chemical disintegration of the material is set in motion. The critical temperature lies above 392° F (200° C) for thermoplastics. If the material is only moderately heated, so that it only softens and becomes flowable, these plastics can be molded and reshaped again and again.

Thermoplastics can be welded

The heat-dependent properties of thermoplastics also enable us to weld these materials. Contrary to the welding of metals, the plastic material is not liquefied in the melting zone and does not melt entirely. The material should be heated only to the point where it becomes doughy. The welding rod is then pressed into the joint. In welding films or foils, the two films are put one on top of another and welded by simultaneous pressing and heating. The principle of welding thermoplastics is the tangling of the threadlike molecules in the welding zone. The welding of thick thermoplastic parts is a rather tricky job, requiring great skill and special tools, and, therefore, unlikely to be of much interest for the hobbyist. There are other films or foils that can be welded with a normal electric iron, which is available in any household, or with a special film-welding unit, which is on sale for sealing polythene bags for deep-frozen food.

The properties of thermoplastics are governed by the structure of their molecular building bricks (monomers) and by their composition within the macromolecule. Another way of influencing the properties of thermoplastic materials is by adding certain fillers. In order to increase their rigidity and strength, fine, short, cut glass fibers, for instance, can be added to thermoplastic injection compounds. This is a common practice in the construction of casings and similar structural parts.

Expanded Thermoplastics

Besides massive thermoplastics, there are also foamed plastomers, which are mainly used for the packing of fragile goods and for insulation purposes. Well-known examples are expanded polystyrene, which is available in a bright white color and in pastel shades and, at a closer look, clearly reveals its granulate structure, and PVC foam, which is sold in thick sheets having a yellowish color. Its porous structure is made up of thousands of tiny bubbles.

Expanded polystyrene is most frequently used for packing, as it can be easily molded into protective shells that snugly fit the goods to be packed. To produce such shells, a pre-expanded granulate, containing a blowing agent activated by heat, is introduced into a metal mold that is filled with hot steam. The plastic material softens, while the blowing agent contained in it is set free and expands the little beads. Due to their increase in volume, the beads are pressed against the walls of the mold and against each other. They fill the mold completely, and the temperature joins them together to form an integral molding. Quite often, expanded polystyrene is called styrofoam. In the German-speaking countries, this foam is often called "Styropor." This, however, is not a general term for expanded polystyrene, but a registered trademark for expandable polystyrene types of BASF, the company that first developed expanded polystyrene. Polystyrene foam can be easily shaped and is in popular use by hobbyists, model makers and craftsmen. It is even possible to build full-size furniture from this interesting material.

DUROPLASTICS

The second large group of plastics are the duroplastics. Once hardened, they cannot be reshaped by heating without distortion. Duroplastics—which are also known as duromers, according to the new nomenclature—do not melt, because their molecular structure—unlike that of thermoplastics—is not held together by merely physical force. When the resin is hardened, duroplasts form a very strong and rigid network of three-dimensional chemically interlaced macromolecules. This is why the hardening of duroplasts is called "interlacing" and why the hardener of liquid plastics is also called "interlacing agent."

Under ideal conditions, a duroplast cast consists of one single macromolecule, which is so tightly and firmly interlaced by threadlike molecular bridges forming the three-dimensional network that the molecules cannot be energized to oscillate under the influence of heat, as the molecules of thermoplasts do.

Properties and Working Methods

Thanks to the very stable netlike structure of the macromolecule(s) of duroplasts, their mechanical properties are only slightly influenced by temperature. In most cases, duromers are brittle and hard. But they can also be elasticized or their hardness can be controlled by a special choice of their molecular constituents and by the degree of interlacing (in other words, by the amount of hardener). Polyurethane (PU) plastics provide a most impressive example for this technique, as their properties can be easily controlled between a glasslike brittleness and a rubberlike elasticity, so that they are something like a bridge between duromers and elastomers. But first of all, we should have a closer look at the duroplasts.

Their brittleness can be controlled by the addition of fillers. Suitable additives are ground minerals, wood flour or wood chips and fibers in the shape of chopped-strand mat or scrim strands, short chopped fibers or fiber cloth or tissue. Glass fibers are a widely used reinforcing material giving the duroplasts a good tensile strength, rigidity and a better resistance to shocks and impacts.

We meet duroplasts in two different states: as molded parts (industrial products from molding compounds that solidify under heat and pressure) and as liquid resins, which are used not only in industry but also by craftsmen and hobbyists who can make their own casts or improve other materials by coating them with such resins or resin compounds.

Bakelite, the first fully synthetic plastic material, was developed by the Belgian inventor Baekeland in 1907 and is based on coal tar. It is a typical representative of the group of molding compounds that are widely used in industry. Bakelite is produced by molding a mixture of fillers and raw materials in a two-step process. This mixture is molded under pressure in a heated mold. Even today, Bakelite is still in large-scale use for bonding the sand molds used for metal castings and even for such spectacular items as heat shields for spacecraft. Molding compounds are also used for casings of electrical equipment, machines and switches.

Bakelite as a heat shield for spaceships

All these materials feature very good mechanical properties and quite a good chemical resistance. They are supplied as semimanufactured or finished products that are quite interesting for do-it-yourself jobs.

Cold-cure Liquid Resins for the Handyman

Cold-cure liquid resins offer a very wide scope to both amateurs and professionals. High-quality casts can be made from such resins by a fairly simple do-it-yourself procedure. Combined with glass fibers, cold-cure liquid resins can be used to make light yet very strong boat hulls or car bodies. But you can also build a small ornamental garden pool or even a full-size swimming pool from these materials. They enable you to mend rust holes in the body of your car or to seal a leaky flat roof once and for all. Epoxy resins in bright and brilliant colors are ideal for enameling work, and they do not require a kiln. Polyester resins mixed with suitable fillers provide decorative tabletops and window benches.

Versatile in combination with glass fibers

Easy-to-use duromer foams make boats unsinkable; insulate against heat, cold and noise; reinforce lightweight structures or protect the cavities of car bodies and chassis against corrosion from inside. Leaky wooden boats can again become watertight and can be given additional structural strength merely by a coat of resin reinforced with glass fiber. Resins mixed with metal powder can be used for repairing faulty metal parts or for sealing leaky containers.

The wide range of applications and the versatility of liquid plastic resins, which include lacquers, protective coatings and sealing agents—all hardened by the addition of catalysts—open a wide and most interesting field of activities to the handyman.

For this reason, an extra large part of this book deals with this kind of plastics.

ELASTOMERS

The third group of plastics are the so-called elastomers. They, too, are characterized by a three-dimensional network formed by their extended molecules. Just as with duromers, the

Elastomers are as elastic as rubber

links or knots of the molecules of the elastomers are of a chemical nature, but the molecular mesh is much wider than that of duromers. Because of the greater distance between the single knots of the net, elastomers are as elastic as rubber. In classifying plastic materials, the following definition may be useful. Macromolecular materials, which having been stretched at room temperature to at least twice their length will snap back to approximately their original length when the stress is removed, are called elastomers.

Elastomers, like duromers, do not melt at higher temperatures, but remain elastic, like rubber, until their molecular structure is broken up by chemical disintegration. There are many ways for the practical man to use elastomeric plastics. They serve

Mold-making material, joint fillers, embedding material

as an ideal mold-making material for intricate castings, where either silicone rubber, PU compounds or even latex can be used. Silicones and PU compounds are also most suitable for filling materials, as both seal joints reliably. Electrical and electronic circuits can be embedded in silicone rubber (just as in epoxy resin). Besides massive, nonporous elastomers, there are also expanded types. This group includes soft PU foam, which can be bought in blocks in department stores and is an ideal material for homemade cushions, mattresses, easy chairs and sofas. PU foam is also available as pourable foam, which belongs to the category of liquid plastic resins. Within the scope of the above applications, elastomers are useful for a large number of repairs and construction work for both the amateur and the professional.

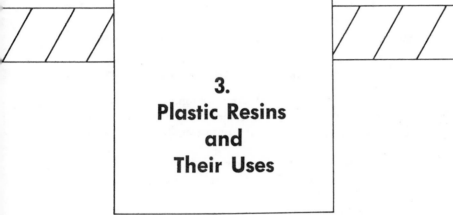

3.
Plastic Resins
and
Their Uses

The family of liquid plastic resins is widely branched and includes both duromers and elastomers. Unsaturated polyester resins, epoxy resins, PU resins and silicone rubber are the most important for the amateur craftsman. Polyester and epoxy resins and some of the PU resins belong to the group of duromeric plastics. Silicone rubber and some polyurethanes are of the elastomer group. Each of these four groups of resins is available in a wide variety of types, with special properties that make it possible to use a tailor-made resin for each and every special purpose.

Most plastic resins represent so-called two-component systems. The interlacing (hardening) only takes place after the mixing of the basic component (also called A-component, or, simply, resin) with the interlacing agent (also called B-component, or, simply, hardener). In addition, there are also one-component liquid resins, the chemical hardening of which is blocked by a solvent. Such resins are mainly used as primers and sealing agents and are applied in thin films, like a lacquer or varnish. Once the solvent has volatilized, hardening starts in one-component PU resins; for instance, air moisture acts as a partner in the interlacing reaction.

PLEASE BEAR IN MIND:

Two of the three main groups of plastics are available as liquid resins: the duroplastics and the elastomers.

Liquid plastic resins are available as one-component and as two-component systems.

ONE-COMPONENT LIQUID PLASTIC RESINS

As easy to apply as lacquer

Due to their special composition, one-component liquid resins are exceptionally easy to use. In practice, they are applied just like a lacquer and are ready to use after vigorous stirring or shaking.

But there are other features shared by one-component resins and lacquers. Generally "one-can" resins only have a limited shelf life once the container has been opened. Once a can has been opened, the lid may no longer fit tightly, and the solvent may slowly evaporate in storage. After some time, the hardening reaction cannot be stopped. While lacquers generally only form a skin on top, one-component resins turn hard throughout and are useless.

Take care when storing resin

Special care should be taken when storing one-component plastic resins that react with humidity from the air. The layer of air above the liquid in a partly used can may contain enough moisture to make the resin gel. Therefore, it is worthwhile to buy the required resin for bigger jobs, which can only be done in parts, in several small cans instead of buying one big container, which then must be stored in a partly used state over a fairly long period.

The label on the cans should always indicate the shelf life (i.e., the time the material can be stored without any risk), and storage conditions. If there is no such indication, ask the dealer or manufacturer.

Only apply thin coats

Again, like commercial lacquers, one-component resins should not be applied in a single thick coat but in fairly thin films to ensure a proper hardening. If this point is neglected, the plastic coat will cure on the surface and block the evaporation of the solvents contained in the still liquid material underneath. Just as with enamels applied too thickly, the resin will remain soft and doughy and tend to form bubbles when the solvent underneath starts to evaporate and rises toward the surface.

Finally, remember to use only specified thinners for plastic resins. You should also refrain from mixing your material with another brand; to do so is to risk failure. The correct thinner is generally indicated by the manufacturer.

Use the proper thinners

TWO-COMPONENT SYSTEMS

Unlike the one-component liquid resins, which are supplied ready for use, two-component systems require some preparation. Some specific rules have to be followed when working with two- or more component resins. People who prefer a slapdash approach and do not much care for the mixing proportions and directions for use must resign themselves to possible failure. It pays to follow the instructions on the label—those who do so will achieve good results and work with liquid plastic resins again and again.

The Right Proportions

Bear in mind that two-component plastics consist of a resin component and a hardener, which transforms the reactive resin molecules into a three-dimensionally braced and interlaced molecular structure that forms a rigid or elastic plastic material. It will then be obvious that resin and hardener must be mixed in a certain proportion in order to achieve the desired molecular structure. (The hardener is sometimes called the catalyst.)

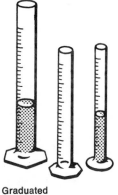

The permissible margin in the proportion of resin and hardener varies with the different types of resin. The more precisely you weigh or measure the components, the more likely you are to achieve a good result. This is why one should never try to prepare a resin mix by guesswork.

Graduated measuring glasses

With some resins, a smaller amount of hardener will only slow down the hardening process, while too much hardener will make the resin gel within only a couple of minutes (e.g., with

Quantity of hardener

polyester resins). With other types of resin (e.g., PU resins) too little hardener will have a permanent effect on the properties of the hardened plastics. The expert talks of "undercuring," where some of the interlacing links that normally brace the molecular structure are missing; vice versa, too much hardener makes the molecule too rigid, and a normally flexible material will turn out to be as brittle as glass due to overcuring.

The exact mixing proportion of both components is always indicated on the labels of the cans and refers to the *weight* of the components. With very thin, liquid hardeners, however, the quantity of hardener to be added may also be indicated by volume, while the quantity of resin required is based on the weight.

FOR EXAMPLE: Add 7 fl. oz. (7 cc.) of hardener to 2 lb. 3 oz. (1 kg.) of resin or, if a dropper is provided, add 20 drops of hardener to 3½ oz. (100 g.) of resin.

Accelerators

To start the reaction, an accelerator is generally required in addition to the hardener. This accelerator is often already added to the resin by the manufacturer for easier handling of the resin. The quantity of accelerator added controls the speed of the hardening reaction, i.e., the so-called curing time. But you can also buy resins without the accelerator, so that you have to add it yourself before the hardener.

Most accelerators are very active substances, which are only added in very small quantities. This may be quite difficult for a layman, who generally does not have sufficiently precise measuring equipment.

With small mixes, exact measurement of the required quantity of accelerator is nearly impossible, so that hobbyists generally do better to use resins sold already mixed with accelerator ("pre-accelerated" resins).

If larger quantities of resin are used and if it seems advisable to control the curing time for technical or practical reasons, one can also use resins that are not yet accelerated. In this case, however, one should already have some experience in the use of liquid plastic resins. The addition of more accelerator to pre-accelerated resin is a further possibility; this is advisable if more rapid curing is required or if the resin has been stored for a fairly long time. Accelerator may partly decompose in storage and lose some of its power. In such a case, it is generally necessary to check how much the hardening is retarded with the normal quantity of hardener added and then to make several tests with slightly increased dosages of additional accelerator. Add more and more accelerator, step by step, until the normal curing time is achieved. Always mark down the additional quantity of accelerator used in each test mix, as well as the curing time. This will enable you to calculate how much accelerator will be required for a certain quantity of resin you need for a special purpose, in order to achieve the original "pot life." (Pot life means the interval between the end of mixing resin and hardener and the point when the mixture starts to gel. This is also the period in which the reactive material can be worked. Please bear in mind that the temperature too has some effect on pot life.)

Pot life

Graduated beakers

Stirring and Mixing

Some resins tend to form a slight deposit during storage. With resins that are mixed with fillers, the relatively heavy additives will tend to settle on the bottom of the container. It may also happen where mixtures of resins with different specific gravity are used that the heavier resins will settle. With some other resins, prolonged storage may lead to partial crystalization. This is the reason why it is always necessary to stir mixtures carefully before they are weighed and mixed. Always stir the resin until it is a homogeneous mixture. Another way to make sure that the resin gets back into perfect condition is to transfer it to another container once or twice and to stir it each time. Only after both components have been stirred separately are the corresponding quantities of both components weighed. It is self-evident that the same wooden stirrer must not be used in resin and catalyst, for then the resin or hardener would be unsuitable for further storage. For small mixes, cardboard beakers impregnated with paraffin may be used. Injection-molded polythene bowls of the kind used for mixing plaster are ideal, but cans or plastic buckets will do, providing they are round—corners make mixing difficult.

Plastic mixing bowl

By the way, some manufacturers sell their liquid resins in calibrated containers, which do not require weighing or a special mixing pot. The container is big enough to take the hardener as well and to allow proper mixing. As always, both components must be thoroughly mixed. Do not forget the resin near the walls of the container or in the lower part. Hold the stirrer at an

Wooden stirrer for small polyester mixes

angle to bring the resin from the bottom of the container to the top. It is most important to stir smoothly and steadily and not to introduce too much air into the resin, as this would cause air bubbles and spoil the laminate. The resin should not be whipped like egg white, but mixed like dough.

Some compounds are rather difficult to mix with hardener, for instance, mixtures with a high percentage of filler or large mixes of 25 lb. (10 kg.) or more. These cannot be properly mixed by hand and call for mechanical mixing. An electric drill with a mixing attachment will do the job very well. If the manufacturers specify that the resin should be mixed with a mechanical mixer, it is no use trying to mix it by hand.

Perfect mixing of both components is most easily perceptible if the hardener has a distinct color that contrasts with the resin. Sometimes, even the "feel" during mixing will indicate whether the components are well mixed or not. This is the case if there is a fairly great difference in the fluidity (viscosity) of the resin and the hardener. If the mix produces an even resistance when it is stirred, you may expect that the mix is satisfactory.

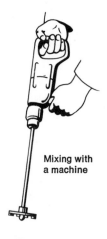

Paste hardeners, by the way, are difficult to mix with very viscous syruplike resins. This difficulty is easily overcome by mixing a small quantity of resin with the entire amount of hardener and then working this premix into the remaining resin.

Even though mixing should be as thorough as possible, in most cases it is not desirable to continue mixing too long, as this shortens the pot life; so, too, the use of a mechanical mixer may be unwise. The energy of friction generated in bringing the required quantities of hardener near to the single monomeric building stones produces a certain amount of heat which accelerates the hardening and reduces pot life and working time. When mixing liquid pour-in-place foams, overlong mixing may even destroy the foam as it develops.

The pot life may also be reduced if fresh resin and hardener are mixed in a pot containing the residue of the last mix, which has begun to cure. The working temperature, too, is important. Heat accelerates curing, and cold retards it. The approximate pot life indicated on the label generally refers to a room temperature of 64.4–68° F (18–20° C).

Even the quantity of resin mixed in one batch affects the pot life. Large mixes will cure more quickly than small ones, because the heat caused by the reaction accumulates in the mix (poor conducting power), and the quickly rising temperature accelerates the hardening process. Liquid plastic resins should only be mixed in clean receptacles, so as to prevent the influence of any foreign substances. Do not add thinners to resin, except those (if any) specified by the manufacturer. As solvents evaporate and thereby cause a reduction in volume, allowance must be made for increased shrinkage during curing if extra solvents are added.

Big quantities harden more quickly

A certain amount of shrinkage will take place anyway during the hardening of some resins. It may be up to 8 percent by volume (with pure, unsaturated polyester resin), but with some resins (epoxy resins and PU systems), shrinkage is negligible, and they are, in fact, known as shrinkage free. The "natural" shrinkage during hardening is due to the composition of the huge molecule, which consists of many monomeric building bricks. Depending on the nature of the interlacing of the macromolecule (polymerization or polyaddition),* i.e., the way the molecules of the resin react with the hardener, it causes the newly formed macromolecule to be shorter to a greater or lesser degree than the total length of the molecular constituents before interlacing.

Shrinkage up to 8 percent

As the pot life puts a strict limit on the quantity of activated resin that can be used without running into trouble, it is advisable not to make up too much at one time. Any resin that turns hard is completely wasted. When deciding how much resin to mix, please bear in mind that liquid plastic resins can be removed when gelling but not once they have completely gelled or cured. It is necessary, therefore, to consider not only the time for working the resin but also for cleaning of tools.

Once set into motion, the reaction of resin and hardener practically dictates the speed and sequence of your operations, so that careful and deliberate planning is essential. All the tools and gadgets required and, above all, the suitable solvents must be at hand. Moreover, all preparations should be finished when the resin is mixed with hardener. Before starting up the "chemical witchcraft," you will find it useful to think over all the operations once again. Missing materials can then still be fetched,

Rule Number One: careful planning

* See, for instance, "polyester" and "epoxy resin" on pages 27–28.

and if you have overlooked one of the necessary preparations, you can still make up for it.

Then, you may confidently begin to mix the resin and hardener. Any negligence will certainly find you out—time and pot life wait for no man. Once activated, the resin will gel remorselessly, undeterred by the most unprintable language!

Safety Precautions

As a matter of fact, the use of liquid plastic resins generally does not involve any greater risk or danger than work with conventional materials. A few precautions will guard against any kind of possible danger.

No smoking! Inveterate smokers should take a few minutes off now and again to smoke a cigarette outside the workshop but do without their beloved cigarette or pipe while working with resin. Many liquid plastic resins are highly flammable. This also applies to most of the indispensable solvents and cleaning agents.

Ensure good ventilation The vapors of solvents are especially dangerous as some of them are heavier than air, so they can settle on the bottom of the workroom without being noticed. If they are flammable or even explosive, like the vapors of acetone, the standard cleaning agent, they may even be ignited by a flame that is quite far away from the actual working site and start a fire. Another danger may arise from the displacement of the air in the room by solvent vapors, which may lead to deficiency of oxygen for breathing.

Caution: caustic and toxic. Keep children away! Plastic resins and their components may be toxic or highly caustic and should be stored out of the reach of children. Do not allow chemicals to come into contact with your mouth, eyes, nose or open wounds.

People who are subject to allergies or asthma should take special care, or better still, ask their doctor if there are any objections to their working with resin. The best possible ventilation—or, even better, working in the open air—and protective plastic gloves are indispensable precautions in such cases.

Before working with a new material, it is worthwhile to read the instructions most carefully and to pay special attention to any safety rules. Then, you can be sure that your working with liquid plastic resins will have no alarming consequences.

In the following chapters, the description of each group of resins is introduced by a brief "record," which includes hints of possible dangers.

Any general points to be borne in mind when working with two-component plastics are included in the synopsis below.

CONCISE LIST OF HINTS FOR THE GENERAL HANDLING OF TWO-COMPONENT PLASTICS

PURCHASE: Quote the purpose for which the resin is going to be used, consider the shelf life of the resin, do not order excessive quantities.

STORAGE: Generally dry and cool (50–64.4° F = 10–18° C). Preparations before use: If necessary, stir thoroughly or shake the bottle or can. When opening a container, never aim its opening at your own or other people's faces, as the container might be slightly pressurized. Danger of splashes!

PROPORTIONS: Carefully adhere to the indicated mixing proportions, weigh the components or measure the volume in calibrated containers.

MIXING: Follow the mixing instructions (by hand or with a machine), observe the mixing time, mix thoroughly and do not stir air into the resin. Allow the mixing tool to come to a standstill while it is still immersed in the resin in order to avoid splashes. Only mix as much resin as you can use up within the pot life without trouble or hurry.

FURTHER USE: According to the specifications of the manufacturer.

GENERAL HINTS: Keep cleaning agents and some rags at hand. Carefully seal partly used cans immediately and turn them upside down to seal around the lid. But store them right side up and, if possible, mark the date when the can was first opened on the label.

SAFETY PRECAUTIONS: Avoid fire, smoke and naked light bulbs, ensure good ventilation. If you have sensitive skin, wear protective gloves or use a barrier cream. Wear goggles when working with caustic materials. Keep chemicals out of reach of children. Dispose of any waste in accordance with the manufacturers' recommendations.

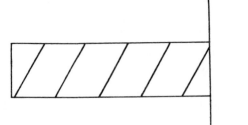

4.
Synthetic Resins

UNSATURATED POLYESTER RESINS

USES: in combination with glass fibers for casts • for the coating of wood, metal or concrete • for repairing corroded metal (rust on car bodies) • clear resin for embedding with fillers to cover cracks and for sculpture • with metal or ceramic powder, for metal or ceramic repairs • with chopped glass fibers for lamination and reinforcement • as foam for light construction work.

CHARACTERISTICS: strong smell similar to that of coal gas, honey or hyacinths.

CAUTION: Resin and hardener and accelerator are flammable! Hardener caustic!
    Never mix hardener and cobalt accelerator, as the mixture is explosive.
    Ensure good ventilation, especially if large quantities are used!

Versatile Material

As you may see from the uses listed, unsaturated polyester resin is a most versatile material. There are many special types and special modifications of resin, with special properties for specialized applications. There also are two different hardening systems—some resins require a liquid and some a paste catalyst. Liquid hardener (MEKP = methylethyl ketone peroxide) can be most easily mixed with unfilled resins, which are used in combination with glass mat or woven glass fibers for casts, while the paste is widely used for polyester fillers and special resins for car body repairs. When mixing the paste with liquid resin, it is advisable to mix first a small quantity of resin with the hardening paste and then stir this premix into the remaining resin until it is evenly dispersed.

Liquid or paste hardener

Just as there are two different hardening systems, there also are two different ways of controlling the cure rate by the addition of accelerators. Besides the liquid dark violet-blue cobalt accelerator, there are also various amine accelerators, which, however, only have an effect on the pastelike Bp hardeners (Bp = benzoyl peroxide), while cobalt accelerator can be used for both types of hardener.

Cobalt or amine accelerators

As MEKP hardener mixes more readily, and, in combination with cobalt accelerator, has a wider margin of error, this system is widely used for laminated parts used in industry. The temperature must not be allowed to fall below 53.6° F (12° C). At lower temperatures, curing is inhibited. If the temperature then rises, setting can be expected to recommence. It is based on the chemical disintegration of the oxygen-rich hardener by the action of the accelerator. The oxygen set free by the disintegration of the hardener reacts with the resin (polymerization).

Minimum temperature 53.6° F (12° C)

If an urgent repair necessitates the use of polyester resin at a temperature below 53.6° F (12° C), you can use one of the tricks of the trade. The solution of this problem is a so-called double- or cross-hardening: both liquid MEKP hardener combined with cobalt accelerator, previously added to the resin, and Bp paste hardener, in conjunction with amine catalysts, added to the resin. This allows working at a temperature of 32° F (0° C) without any risks. A pre-accelerated standard

Cross-hardening at low temperatures

resin* (containing 0.2 percent of cobalt accelerator) has to be mixed with an additional 1 percent of cobalt accelerator and 1 percent of highly effective dimethylparatocuidin (DMPT) amine accelerator and, finally, 3 percent each of MEKP and Bp hardener.

Light Aids Polyester Curing

The great variety of special types of resin includes the ultraviolet-sensitized polyester resins. They contain optically sensitive substances that oscillate in a sort of sympathetic vibration if exposed to light of a special wave length, and this causes the liquid resin to cure. In industrial production, such resins are used for polyester-finished furniture. In this case, the resin-coated surfaces are passed through a hardening tunnel that is equipped with high-pressure mercury vapor lamps, which harden the resin and provide a colorless clear film within a very short time.

Such resins are also available for the hobbyist who can use them for quite interesting and versatile experiments. There is no need for special lamps to start the curing of the resin, for even the light of the sun gels these resins. The practical application of ultraviolet-hardening resins is more limited than that of resins using added hardener, although there may be further developments. It is already possible to produce resin-impregnated glass fiber material that can be put on surfaces to be coated or repaired and will harden by sunlight.

Characteristic Properties

As already mentioned, polyester resin is subject to shrinkage up to a maximum of approximately 10 percent (unfilled resins). Moreover, cured polyester is very brittle. Both these undesirable characteristics can be controlled by adding fillers to the resin. In combination with glass mat or woven glass fibers, the shrinkage of polyester resin is reduced to about 2 percent, and, at the same time, the material becomes remarkably resilient. This composite material is quite within the reach of the amateur. The

Elasticity
by fillers and
glass fibers

* This formula relates to the standard resin of Vosschemie and Bondaglass-Voss Ltd., which has a pot life of approximately forty minutes.

POLYMERIZATION OF POLYESTER RESIN

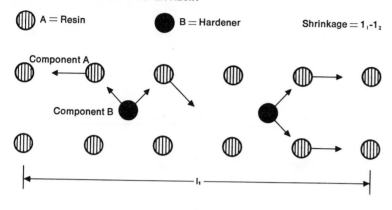

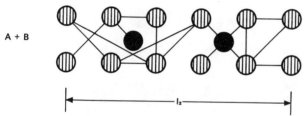

With the hardening of polyester resin, the hardener finds many starting points for polymerization, which means fast hardening accompanied by shrinkage (l_1-l_2 = shrinkage).

section "The Practical Use of Liquid Plastic Resins—Laminated Casts of Glass and Resin" (pages 32–41) describes the method.

Polyester resin is a relatively inexpensive material, which helps to explain its wide use. Two characteristics that at first might appear unfavorable prove to be quite advantageous in practice. The slight shrinkage, which takes place when the resin is used in combination with glass fiber, is just enough for the cast to "shrink" itself loose from the mold, which makes removal easy. Even if shrinkage leads to adhesion problems, as with glass fiber coatings, for instance, there usually are means of overcoming the problem. There are, for instance, special resins for car body repairs featuring a minimum shrinkage and sticking firmly to the metal, owing to special additives that improve adherence. There are, however, some materials with a very smooth

Shrinkage

Car repairs

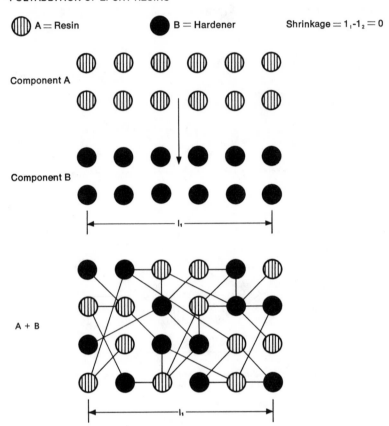

・POLYADDITION OF EPOXY RESINS

A = Resin   B = Hardener   Shrinkage = $1_1\text{-}1_2 = 0$

Component A

Component B

A + B

Epoxy resins harden by polyaddition in a fairly slow process without shrinkage.

and dense surface, such as glass, where even these special types of resin must fail. On the other hand, such materials can be used to advantage wherever adhesion is not desired. Glass panes, as well as laminated plastics, such as Formica, Micarta, and chipboard with a melamine finish, make quite good molds for casts that are flat on one or both sides.

Another property that seems unfavorable at first sight turns out to be an advantage when the resin is used with glass fiber for laminating.

Laminating: tacky surface   Standard polyester resin does not completely harden on the surface that is exposed to the air but remains slightly tacky. This very thin, tacky film can be removed by washing the surface with solvents, or better, by giving it a sealing coat with a special polyester lacquer that hardens with a nontacky surface. The reason for the formation of the tacky film is that the monomeric styrene, which makes up about 30 percent of the polyester resin, volatizes before the hardening takes place, so that there is no chemical reaction on the immediate surface of the resin. If this tacky film is wetted with a fresh mixture of resin and hardener, it will completely cure together with the new coat of resin, which, however, produces a new tacky film of its own on exposure to air.

This enables us to interrupt our laminating work for a break of a few hours, or even days, without running the risk of getting a poor adhesion between the single layers of resin-impregnated mat. On the other hand, it is also possible to avoid the tacky film if it is troublesome. In such a case, you have only to seal the surface of the fresh laminate with a film or foil that is impermeable to air. You can use oiled paper, which can be easily removed once the resin has turned hard, or Mylar or Melinex film or foil, to which hardened polyester resin does not adhere.

This trick is also used with the so-called polyester LT resins, (LT = *lufttrocknend*, the German word for air drying), which produce a nontacky surface. The sealing coat that prevents the styrene from volatilizing into the air is formed by paraffin wax dissolved in a solvent. It settles on the surface of the resin applied to the laminate and forms a sealing barrier. Unfortunately, this only works at temperatures above 59° F (15° C), as the resin will not cure at a lower temperature, and it always produces a slightly dull finish. But high temperatures are difficult, too, as they make the paraffin wax melt and run off, which again leads to a tacky surface. Such mishaps can only be cured by applying a new coat of resin when the temperature is above 59° F (15° C) and below 77° F (25° C).

High temperatures melt paraffin wax

Quantity of Hardener—Curing Time—Final Strength

The question of how much hardener has to be added cannot be answered by giving a precise figure. The amount generally ranges between 1.5 percent and 3 percent. This applies both to liquid and paste hardener and refers to the weight of the resin. With very large mixes and warm weather (about 86° F = 30° C), less hardener will do. With small mixes and cool working conditions, 3 percent is the standard rule. If you are in a hurry, you may add up to 5 percent of hardener, but this will not lead to an increase of either strength or hardness of the cured resin.

Even more hardener will excessively heat up the curing reaction, which produces heat anyway, make the resin extremely brittle and may even lead to tension cracks. With polyester air-drying resins, too much hardener may even prevent curing altogether. Although this may seem surprising, the effect can be easily explained, as the highly reactive MEKP hardener is generally not used in its pure form but mixed with plasticizers, namely nonvolatilizing solvents. They are added as a safety precaution in order to stabilize the hardener. These plasticizers, which are also used for making thermoplastics flexible, keep the resin soft if too much of the stabilized MEKP hardener is used. The plasticizers contained in the hardener will then "sabotage" the hardening action.

Pot life of about forty to fifty minutes

As already mentioned, the pot life and the curing time very much depend on the temperature. At normal room temperature and with the normal quantity of hardener added (2 percent to 3 percent), standard resin will have a pot life of approximately forty to fifty minutes. In this case, the cast can be taken out of its mold after about twenty hours (at 68° F = 20° C curing temperature) or twenty-eight hours at a curing temperature of only 64.4° F (18° C). At a higher curing temperature, only 95° F (35° C), the cast may be taken out of its mold after only two hours, without fear of warping. Even though the cast seems to be completely hard when removed from the mold, hardening continues for some time after that. If extra high strength is desirable, the cast should undergo a final heat treatment after removal from its mold. This may be done in a water bath or even in the heating chamber of a car body repair shop. By heating the cast to a temperature of 140° F (60° C) for about one hour, the cast will undergo a noticeable further hardening. At 176° F (80° C), this effect can be reached in less than half the time.

EPOXY RESINS

USES: in combination with glass fibers for high-strength casts • for coating metal and concrete • for embedding electronic circuits • for making high-precision templates and patterns of high dimensional accuracy • as adhesive for metal, glass and other nonporous materials • filled with metal powder as a first-class molding and repairing compound for metal parts.

CHARACTERISTICS: generally rather viscous, similar to concentrated syrup; odor and color may vary.

CAUTION: Hardeners are toxic! Prolonged contact with the skin may involve toxic reaction! Always wash your hands with plenty of water and soap. Ensure good ventilation in your working room.

Any activated resin left over must be poured away over a large area; with some types, heat accumulation may lead to spontaneous ignition.

More Expensive, but Sometimes Better

Specialized resins do a better job, but unfortunately they are also more expensive. This also applies to epoxy resins. They cost about six times as much as polyester resins. Epoxy resins cure by polyaddition virtually without shrinkage and perfectly adhere to nearly everything, including metal and glass. Their resistance to many chemicals is higher than that of polyester resins. They can be used for fuel tanks for motor bikes or boats without fear of chemical reaction and its dire consequences when particles of resin are burnt together with the fuel.

Epoxy resins cure almost without shrinkage

A further advantage is the lack of stress during the hardening process, which is most useful with regard to the embedding of rather critical electronic parts. As such parts, however, are very sensitive to temperature, care must be taken that the type of resin used does not generate too much heat when hardening and that the casting is built up in sufficiently thin layers in order to avoid damage by heat accumulation.

But there are not only advantages. One of the undesirable disadvantages is the high price of epoxy resins already mentioned. A further drawback is the smaller margin of error in the proportion of resin and catalyst, with some as little as 1 percent, although other types have a margin of about 5 percent.

Disadvantages

In most cases, however, even a deviation of plus or minus 5 percent will change the properties of the hardened resin.

Last, but not least, the higher viscosity of epoxy resin is an impediment. The resin does not flow as easily when it is used for casting purposes, and when the resin is used for laminating, the impregnating of the glass mat is much more difficult than with the much less viscous polyester resin.

Finally, it should be stressed that the hardeners are toxic, which may lead to a severe headache if large quantities of resin are used in a room with poor ventilation. Therefore, open the window when working with epoxy resin!

The hobbyist should carefully consider the pros and cons before deciding to use epoxy resin rather than polyester resin.

Generally, the decision will be in favor of epoxy resin if special chemical or mechanical requirements have to be met and shrinkage cannot be tolerated.

In modern techniques, epoxy resins are also widely used as glues. This is a field where they also are very interesting for both amateurs and craftsmen.

20 percent to 100 percent of hardener

Contrary to polyester resins, the amount of hardener to be added to epoxy resins is much bigger. It depends on the type of resin, and amounts to between 20 percent and 100 percent in relation to the quantity of resin. The pot life of standard epoxy resins is generally only a little longer than that of polyester resins. As a result, the time that must pass before the casts can be taken out of the mold is a bit longer, too. As there is no shrinkage, the casts are not so easy to remove from the mold as polyester casts.

THE PRACTICAL USE OF LIQUID PLASTIC RESINS— LAMINATED CASTS OF GLASS AND RESIN

High strength and low weight

Laminating casts from resin and glass fiber no doubt represent the main use of polyester and epoxy resins. The compound material built up from these materials excels by its fairly low weight combined with high resilience and strength. This is why these materials have gained an undisputed position in modern technique within a fairly short time wherever low weight and high strength are required together. Another point in favor of this material—which is often called "Fiberglass"—is the fairly simple way of working with resin and glass fiber, which makes small batches economical in industry and opens a wide field of activities to the hobbyist, do-it-yourself people, artists and craftsmen. There are quite astonishing and remarkable examples of the use of this revolutionary combination in many fields of industry and research that bear witness to the outstanding properties of this material. Radar cones and fuselage fairings, or even complete tailplanes or fins of civil or military aircraft doing more than 600 mph, the bodies of nearly all racing cars, the glass fiber pole, which due to its flexibility, catapults the pole-jumper over heights that had never been reached before—people called it a "magic pole" when it appeared for the first time—highly stressed fuel tanks of spacecraft and even fairings of the American Saturn–rocket, which carried the first human beings to the moon, and last, but not least, modern tanks for fuel oil. All of these items are made from glass-fiber reinforced liquid plastic resins, whereas epoxy resins predominate whenever extreme demands have to be met.

The Technique of Reinforcing Plastics—Just a Bit of Theory

The structure of reinforced plastics can be roughly compared with steel-reinforced concrete, where the actual building material, "concrete," is reinforced with embedded steel rods or steel mats in order to increase the tensile strength of the finished product. As with reinforced concrete, the reinforcing of polyester or epoxy resin can be achieved by single strands, which can be compared with reinforcing steel rods, as well as with mats or woven fabrics. The special kind of reinforcing depends on the stress the part has to take. If compared with reinforced concrete, the amount of reinforcing material, i.e., glass fiber, is much greater in glass-fiber reinforced plastics than the content of steel in reinforced concrete.

Strands, mats or woven fabrics

The Relation of Glass Fiber and Resin

In a glass fiber cast, the content of glass may amount to between 30 percent and 70 percent of the total weight. The higher the percentage of glass, the greater the tensile strength, bending strength and impact strength of the laminate. They are, so to speak, the dowry of the glass fibers, contracting a marriage with the fairly brittle resin, which is not very strong on its own.

The glass is used in the shape of very fine fibers, to exploit a special property of glass: the thinner the fibers, the greater the tensile strength. For example, a strand of glass fibers having a total cross section of 4/100 sq. in. (1 mm.^{2}) with each fiber measuring 7/1,000 mm. in diameter can take a tensile stress of approximately 485 lb. (220 kg.), while a second strand with the same cross section of 4/100 sq. in. (1 mm.^{2}), but with each fiber only measuring 2/1,000 mm. in diameter, can take three times this load. This remarkable tensile strength has its drawbacks, however, for glass fibers have poor form-retaining properties and, above all, have very little resistance to abrasion. Rigidity and insulation are provided by the resin in the laminate.

Depending on the proportion of resin to laminate, the specific properties of one or the other predominate. Casts with a high percentage of resin feature high form-retaining properties, but will be rather brittle, with poor impact strength. With a greater percentage of glass fiber, the cast becomes more resilient, so that it can take impact stress without damage. If a laminate is required to have the same stiffness and form-retaining property as the above cast but without its brittleness, one can either increase the thickness of the laminate or—what is much simpler and also, in addition, saves weight and material—reinforce the cast with stiffeners laminated on the back. This is where a construction to suit the material is of real importance.

Before discussing ways of reinforcing fiber laminates and the practical aspects of working with this material, we should take a closer look at the different types of glass fiber products available.

Different Types of Glass Fiber Mat

Fibers and quantity of fibers

The most important and most widely used form is the chopped-strand mat. It has a feltlike structure and consists of 2-in. (approximately 50-mm.) long thin strands of glass fiber, with a diameter of approximately 10/1,000 mm. per fiber. There is also a glass fiber tissue or surface mat that is translucent and of very fine texture. The split-strand fibers produce a greater number of crossing points between the innumerable short cut fibers and thus provide a higher strength, because the resin-impregnated mat forms, so to speak, a greater number of junctions, similar to those occurring in structural steel work.

A smoother and more even distribution of the fibers is most advantageous with thin glass fiber coatings and for the molding of highly stressed small parts, such as fuselages for model aircraft.

Standard random weave mat, which is generally used for molding bigger parts, is of the chopped-strand type. If you hold this mat against the light, you can clearly see its more uneven distribution of the fibers. Here and there, it is completely opaque. Some spots are translucent, and others are nearly transparent. These small areas of structural variation have little effect on the strength of the multilayer casts.

Chopped-strand mat

The individual fibers of chopped-strand glass mat are bonded together by a special binding agent, providing the characteristic feltlike structure of the mat. The binding agent is necessary to keep the mat in shape during transportation and handling. Once the mat is incorporated in a laminate, the binding agent is no longer required, as the resin will now take over the role of the binding agent. As a matter of fact, the binding agent is even unwelcome during the laminating process, as it makes the mat rather stiff. This is the reason why the binding agent must be soluble in the resin. It takes about two minutes until the binding agent is dissolved by the resin. The fibers then become flexible and can be easily molded into the desired shape by dabbing the mat with a brush soaked with activated resin or by rolling it with a resin-impregnated roller to make the fibers fit the contours of the mold. Depending on the type and make of the mat used, the binding agent may dissolve more or less easily, so that there may be differences in the handling of the mat during the laminating process, even though the overall quality of two different makes is identical.

Fibers held in place by a binding agent

In order to achieve a firm contact between glass fiber and resin, glass fiber products for reinforcing polyester or other cold-cure plastics are treated with a finish. The primer contains two groups of chemicals, one to produce a chemical bonding effect on the surface of the glass fibers, the other to encourage the adhesion of the resin to the primer. As polyester and epoxy resins differ from each other with regard to their adhesion power, the finish of the glass mat as well as that of other glass fiber products (fabrics and strands) used for reinforcing the finish must be specially adapted to the resin. When you are buying glass fiber products, specify whether they are going to be used with polyester resin or epoxy resin. For the best results, you have to use the right resin with the right glass fiber.

Primers

When building up laminates of several plies, it is useful to cut the mats roughly to their required size and then tear or comb the edges to make them fray out. This produces a smooth chamfered joint that is hardly visible and which can be completely concealed with a little polyester filler. If the edges are cut, the joints turn out as clearly visible seams, which require much sanding and filler to make them disappear.

Different weights of mats

For the practical man, the weight of the mat—in addition to the type of its binding agent—is quite important. It is quoted in relation to a sq. ft. of mat (or m.²). The most widely used weights of mat are 1 oz./sq. ft. (300 g./m.²), 1.5 oz./sq. ft. (450 g./m.²) and 2 oz./sq. ft. (600 g./m.²). The medium-weight type is generally called "standard mat" and represents the most widely used type. It has excellent handling and molding characteristics and can be recommended to the amateur as his standard reinforcing material. Impregnated with standard polyester resin, this type of mat proves to be amazingly strong after hardening (see table). The weight also indicates the quantity of resin required for impregnating one layer of mat according to regulations; and, in addition, gives the resulting thickness of each layer. The allowance for the resin required is changed by the viscosity of the resin, which varies greatly depending on the working temperature. The higher the temperature, the more viscous the resin, which means that the working temperature must not be too low for glass-fiber rich laminates.

WEIGHT OF MAT		LOAD AT RUPTURE*		CONSUMPTION OF RESIN		THICKNESS PER LAYER	
oz./sq. ft.	g./m.²	lb./in. of width	k./cm. of width	oz./sq. ft.	kg./m.²	in.	mm.
1	300	357.5	65	3.25	1.0	1/32	0.8
1.5	400–500	550	100	3.25–4.9	1.0–1.5	3/64	1.2
2	600	742.5	135	5.53	1.7	1/16	1.5

* This figure is a standard. Practical tests with carefully impregnated test strips led to results that nearly doubled the standard load figures. Nevertheless, it is advisable to base your own calculations on the above standard figures when calculating the strength of casts of your own design. This will give you an additional safety margin, even if you do not manage to achieve the best possible laminate.

With many laminates, chopped-strand mat is the only reinforcement needed, for instance, for swimming pools (see chapter "How to Build Your Own Swimming Pool" on pages 87–95), for many kinds of glass fiber coatings (flat roofs, see pages 84–87, boats, see pages 82–84), for modeling jobs and for laminating that need not take very high loads.

Glass Fiber Reinforcements Increase the Strength

Wherever high loads occur or to avoid the risk that a minor flaw may present under stress (with a boat, for instance), it is generally advisable to use a combination of chopped fiber mat and scrim, which are offered in a variety of types. There are textures with equal strength in both directions and others that have a greater tensile strength in one direction.

In order to meet normal requirements, textures with equal tensile strength in both directions will generally do. They are easier to work with for the nonprofessional, as there is no need to apply the woven material always in one and the same direction.

Building up a laminate from glass mat or woven glass fiber calls for a perfect joint between the single plies in order to exploit the full strength of each layer, i.e., to achieve the highest possible total strength. With coarse textures, however, this may present a problem. In order to avoid delamination, i.e., splitting of the laminate into individual layers, coarse textures, such as woven rovings, are, in most cases, embedded between two layers of glass mat.

Which Weave for Which Purpose

Normally, the glass strands consist of many very fine fibers. Generally, one strand of chopped-strand mat contains 204 fibers made from high-quality alkali-free glass, each measuring approximately 1/2,500 in. (1/100 mm.) in diameter. Only a few methods of weaving, of special interest to the amateur and professional, are dealt with here. In addition to using different patterns, the weaving may also vary in tension of the warp or weft.

GLASS FIBER SCRIM is an easy-to-use glass fiber texture with a weight of 0.26 oz./sq. ft. (80 g./m.²). It is mainly used for coating wood and has a tensile strength of 93.5 lb./in. (17 kg./cm.), which, however, is not enough to withstand extreme stress. Due to its moderate thickness, it does not produce a surface hard enough to withstand higher point loads. This must be borne in mind when this material is used for the coating of wooden surfaces. It is most suitable for reinforcing fuselages and angled joints of the wings of model aircraft. But even the surface of a workbench in a hobby workshop gets a sufficiently resistant surface if it is coated with glass fiber canvas.

For model aircraft

• GLASS FIBER TWILL has a weight of 0.54 oz./sq. ft. (167 g./m.²) and can also be used for two-dimensional coatings. It is easy to handle, and, due to its higher tensile strength of 187 lb./in. of width (34 kg./cm.), it is more suitable for coating wooden surfaces exposed to weather and stress. With regard to point loads, however, this material is not very resistant either. If the surface has to withstand higher point loads, you should choose standard mat, only one layer of which will be superior with regard to point loads, as well as with regard to its tensile strength. Therefore, standard mat is the first choice for trailers. Experts also prefer two layers of standard mat for coating a wooden boat.

Makes wood weatherproof. Especially suitable for modeling

Modeling is another field where glass fiber twill is a good choice. It provides gliders and power model aircraft—neither of which have landing gears—with a sufficiently resistant body, and it is also often used for reinforcing the noses of power or radio-controlled models. Model seaplanes, amphibians and model boats are rendered resistant and 100 percent waterproof by a coating of glass fiber twill and polyester resin.

Sometimes, fine glass fiber twill is also used as an intermediate layer between glass fiber mats. In order to take up higher loads, you can also buy a heavier type of twill, weighing 1.33 oz./sq. ft. (410 g./m.²) and having a tensile strength of approximately 450 lb./in. (82 kg./cm.) width.

• WOVEN ROVING FABRIC is the classic reinforcing material for all laminates that have to withstand high loads. This, again, is available in different weights and features a fairly coarse structure, with a characteristic checkerboard mode of weaving. Both warp and weft consist of thick strands of fine parallel running fibers with no appreciable twist. The standard type of woven roving weighs 2.18 oz./sq. ft. (670 g./m.²) and has a tensile strength of 742.5 lb./in. (135 kg./cm.) width in both directions. As already mentioned, such fabrics are generally applied between two layers of mat in order to prevent delaminating. If you really work carefully, you can apply woven roving layer on layer. But this technique should really be reserved for the expert.

Tensile strength 742 lb./in. (135 kg./cm.) of width

For a perfect finish, the layer of reinforcing glass material next to the surface of the laminate should not be of woven roving. A layer of woven roving lying just under the surface of the molding will easily reproduce its typical structure through the gel coat and give the molding a slightly honeycomb structure. This is why it is always recommended to apply a layer of fine glass fiber twill or canvas after the gel coat has been brushed into the female mold and before other glass fiber reinforcements are applied. Such an intermediate layer will act as a sort of barrier and prevent the rather coarse structure of the following layer of woven material from spoiling the finish of the laminate.

• SURFACE MAT. Excellent results have recently been achieved by the use of surface mat fabric. These are very lightweight polyester fiber mats (for instance, Fiberil^R), which only absorb little resin. Thanks to their pliancy, they can be easily applied and hide the structure of the following or preceding layers of mat or woven materials very well. A medium-weight type fabric, weighing approximately 0.11 oz./sq. ft. (3.2 g./m.^{2}) should be used for laminates. When coating surfaces with glass fiber, a lighter type of fabric, weighing approximately 0.6 oz./sq. ft. (20 g./m.^{2}), is applied as a final layer, before the surface is sealed with polyester resin. Surface tissue improves the appearance but does not add strength.

Other Reinforcing Materials

In addition to fabrics and mats, there also are roving strands that can be used for reinforcing polyester and epoxy resin. They can, for instance, be used for winding containers and cylinders over a split male mold. This technique, however, calls for some skill and is much more difficult than laminating with mat or fabrics. There are not too many applications of roving strands for the amateur. Such strands can be used wherever high tensile stress in only one direction has to be met. They can be used for profiles and frames and extended structures exposed to tensile stress. Rotor blades for full-size helicopters are a typical example from the field of engineering. For craftsmen and do-it-yourself people, the building of a garden chair would be a possible application for roving strands. The framework of such a chair can be molded from roving strands in a female mold. The frame is then reinforced with roving strands that are not yet impregnated with resin. Once the structure is completed, these strands, which form the seat and the back of the chair, are soaked with resin thickened with aerosil powder.

Roving strands reinforce polyester and epoxy resins

Carbon Fibers Stiffen Plastics

A few years ago, glass fibers received competition from another reinforcing material. Carbon fibers, first developed in

WIDELY USED TYPES OF WOVEN GLASS FIBER PRODUCTS
AND THEIR CHARACTERISTICS

MODE OF WEAVING AND TYPE	WEIGHT		TENSILE STRENGTH		CONSUMPTION OF RESIN		SPECIAL FEATURES
	oz./sq. ft.	g./m.²	lb./in. of width	kg./cm. of width	oz./sq. ft.	kg./m.²	
Scrim, type 90 070	0.25	80	93.5	17	1.3	0.4	Used for intermediate layers and for coatings of wooden surfaces (not very resistant)
Twill, type 92 110	0.55	167	187	34	1.6	0.5	Suitable for spherical moldings, deep-drawing quality
Twill, type 92 140	1.33	410	451	82	2.3	0.7	Suitable for spherical moldings, deep-drawing quality
Woven roving, type 92 171	2.18	670	742.5	135	2.6–3.25	0.8–1.0	For reinforcing highly stressed parts, prevents cracks from widening under load
Sandwich, type 92 390	3.64	1,120	1,375.0	250	5.19	1.6	Extremely strong woven reinforcing material, difficult to use for beginners

England, excel glass fibers with regard both to strength and to the amazing stiffness they give to plastic moldings. These excellent properties make it possible to mold even such high-stress parts as fan blades for jet engines, which are already made from carbon fiber-reinforced plastics. Such fibers are produced by a controlled heat treatment of synthetic fibers in a special kiln containing a protective gas. Due to this complicated process of manufacture, these fibers are still quite expensive. The minimum price is about $60 (£25) per lb.

As yet, too expensive for general use

Therefore, such fibers will not be widely used for general purposes in the near future. Nevertheless, they have already overcome the barrier limiting their use to highly specialized purposes in industrial techniques. Resin-coated ribs of carbon fibers, for instance, were used to stiffen some high-performance rowing boats of the German Olympic team in 1972.

Loads and the Structure of the Laminate

As already mentioned, the content of glass and special kinds of the glass fiber reinforcement are of great importance for the strength of a glass fiber shell. It is not only the number of reinforcing layers and their individual strength that decide the strength of the laminate, but the sequence of single layers also plays a part.

If a laminate has to withstand high bending stresses, for instance, one side is exposed to tensile stress, while the other side has to withstand compressive stress. Tensile stress is best met by the use of woven roving, while normal glass mat will do against compressive stress.

If you are going to design your own casts from glass fiber and polyester or epoxy resin, you should study similar pieces in order to get an idea of the best way of building up the laminate. You can also obtain reliable hints and data from the companies supplying you with glass fiber and resin.

HOW TO SET ABOUT IT

Making a Cast

After so much theory, which unfortunately is indispensable, we are now going to deal with practical aspects. The use of resin and glass fiber reinforcements is a quite simple process, which is very similar to wallpapering and can be managed by anybody who does not have two left hands.

Two Different Methods of Work: Mold or Coating

Basically, there are two methods of work: the making of casts and the coating of surfaces. Casts are usually made in a

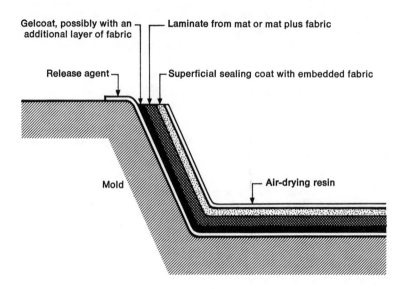

Gelcoat, possibly with an additional layer of fabric

Laminate from mat or mat plus fabric

Release agent

Superficial sealing coat with embedded fabric

Mold

Air-drying resin

Like a cake pan

mold, which you can regard as a cake pan. It is self-evident that such a mold must receive a pretreatment similar to that of a cake pan in order to make sure that the hardened cast will come out without sticking. By using a mold, the cast gets an absolutely smooth surface. It is a prerequisite condition, however, that the surface of the mold has exactly the finish expected of the cast.

The second way of using glass fiber and resin is by surface coating. In this case, it is necessary to achieve perfect adhesion between the coat and the base. Coating may be carried out to improve insulation, resistance or appearance. In most cases, the surfaces are roughened and given a coat of a special primer in order to ensure perfect bonding. The top side of the coating is smoothed by giving it a sealing coat of resin, before it receives a final layer of surface tissue, followed by a coat of polyester air-drying resin to provide the nontacky finish. Both processes are carried out manually by professionals or amateurs.

Laminating in a Mold

This is the ideal method for anybody who works only occasionally with polyester and glass fiber, as it requires a minimum of inexpensive tools and, nevertheless, gives very good results. In most cases, dry patches of mat torn to the required size are laid into the mold or onto the surfaces to be coated in order to be then impregnated with a mixture of resin or hardener applied with a sheepskin roller. The resin dissolves the binding agent and renders the fibers pliable so that they can adapt to the contours of the mold. A change of the color of the mat, which is white and opaque when dry but becomes glasslike and translucent when impregnated with resin, indicates that the binding agent is being dissolved. Light spots indicate air bubbles that must be eliminated. Bubbles with a diameter of more than 1/48 in. (0.5 mm.) reduce the strength of the laminate.

Color changes indicate dissolution of the binding agent

Air bubbles, of course, must be removed before the resin has hardened. Large bubbles can be largely eliminated with the sheepskin roller. The precision work is then done with a metal disc roller, which looks very much like a gadget used in the kitchen for cutting onions or herbs. By rolling, the laminate is squeezed like a sponge, and the resin absorbed by the glass mat rises to the surface with the entrapped air bubbles, which burst, while the resin is reabsorbed by the glass fiber, which then re-expands.

Sheepskin roller for impregnating mats and fabrics

It is useful to work in sections and impregnate about 10 sq. ft. to 15 sq. ft. (1 m. to 1.5 m.²) of glass fiber at a time and lay it up at once. Glass fiber joints should overlap by 1¼ in. to 2 in. Chopped fiber mat should be torn, not cut, as frayed edges make for a smoother joint. With fabrics, however, raised joints cannot be avoided.

Overlap joints

Joints in a laminate should be staggered for an even surface and uniform strength.

Wet into Wet or Wet onto Dry

As already described, when dealing with the properties of polyester resin, standard laminating resin always leaves behind a thin, tacky film that is not completely cured but can be activated when the next layer of glass and resin is applied. It then provides a perfect joint between the first layer, which has already turned hard, and the following fresh one.

Metal disc roller to remove air bubbles

You may also straight away laminate a second ply onto the still fresh and not-yet-cured first one. This procedure is called "working wet into wet." It enables you to bring more glass into the laminate, as it allows you both to apply a bubble-free first layer and to embed it completely, as well as to use the excess resin to impregnate the following layer, using only a little more resin to soak the mat completely.

How to apply the first layer without bubbles

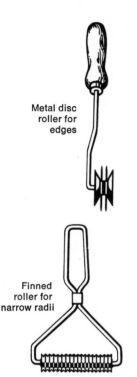

Metal disc roller for edges

Finned roller for narrow radii

This is an advantageous and economic wet-into-wet technique, but, for large laminates, you will need a friend to help you, as one layer has to be completely impregnated before the next one can be applied. With similar parts, however, you can do a wet-into-wet job by yourself. When laminating a cast, all the clear or colored gel coat is brushed into the mold first, and this will form the surface of the cast.

The gelcoat resin is a thixotropic type of resin, which does not run off vertical surfaces. Use a broad, soft brush to apply an approximately 1/64 in. (0.4-mm. to 0.5-mm.) thick coat of resin to the mold, which you have previously treated with release agents. The gelcoat resin is a resilient type of resin. Its special formula prevents the formation of cracks in the surface of the cast when exposed to stresses.

Gelcoat resin must also be mixed with 3 percent of MEKP hardener. As the gelcoat provides the finish of the cast, which should be immaculate, you must never work wet into wet when applying the gelcoat. If you do so, the structure of the mat would imprint the gelcoat and impair the finish of the molding. It is even necessary to allow the gelcoat to harden properly—if possible let it cure overnight—to be sure that the influence of the styrene contained in the laminating resin does not cause the gelcoat to expand.

Impatience causes wrinkles

People in a hurry tend to be left with a poor finish, caused by a gelcoat full of wrinkles requiring much sanding and filling, including even painting, while a sufficiently cured gelcoat produces an immaculate high-gloss finish that does not require any touching up at all.

At one time, a first-barrier ply of very fine glass fiber canvas was used that only absorbed little resin and, thus, reduced the risk of wrinkling. Today, the canvas is replaced by a layer of surfacing mat.

NOTE: Working wet into wet is advantageous from the technical point of view and reduces the consumption of resin. The gelcoat must, however, be allowed to cure completely before you start laminating.

Gelcoat brush

The Mold—Female or Male?

Female mold preferred!

If you want to make glass fiber reinforced casts, you need a mold. In order to achieve a good-looking molding, you should, as far as possible, use a female mold. Working over a core (male mold) seldom leads to satisfactory results and requires a considerable amount of work for smoothing the surface later on. It must be sanded and treated with filler several times, and may even require painting. The application of filler and enamel is a time-consuming job and also causes additional weight, without adding to the strength of the cast. Consequently, the use of male molds is confined to exceptional cases.

The technique has its use for small or individual pieces, such as a prototype of a model aircraft fuselage, which should be laminated in one piece for better structural strength.

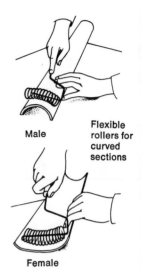

Male

Flexible rollers for curved sections

Female

Dead Cores

A dead core is the best choice where the fuselage is laminated as an integral shell over the core. Any kind of material that can be easily shaped is suitable as a mold-making material, but you should also bear in mind that, in addition, the core must be easily removable from the finished shell. On the other hand, it must also be strong enough so that it does not give way when the glass fiber is applied, and it must be impervious to the resin.

Clever modelers, therefore, use expanded polystyrene foam with a specific gravity of approximately 3 oz./cu. ft. (30 kg./ m.³). This can be bought from a dealer in building supplies or a builder's merchant in the shape of thick sheets or even ready cut to the required size. This material can be easily cut with a hot wire (see page 222) but also, if need be, by means of a 100-w. soldering iron. Very sharp knives allow proper carving if you confine yourself to smallish chips. Finishing can be done with fine sandpaper. Polystyrene foam can even be glued together with white (PVA) glue or special glues. You can also produce your own filler paste from glue and the polystyrene foam dust that is a by-product of the sanding process. This can be used to fill in holes and uneven parts.

A core from polystyrene foam can be cut and carved

A fairly smooth finish can be obtained by coating the core with emulsion paint (PVA paint) or white glue. This coat can be reinforced, if you apply a layer of thin Japanese paper, as used for covering model aircraft wings.

Once this surface coat has turned hard, it can be treated with release agents (see pages 50–52) in order to prepare it for the application of glass fiber and resin, which forms the surface of the cast. If you work with polyester resin, it is advisable to begin with a thin layer of light glass fiber canvas with a little resin and allow it to harden before the actual skin of the molding is built up. This precaution prevents the insulating coat of white glue from being damaged by the polyester resin and, at the same time, stops the polyester resin from penetrating the thin parts of the coating and attacking the foam. The monomer styrene contained in the resin would dissolve the foam in a couple of minutes.

Epoxy resin does not attack the foam

If you use epoxy resin instead of polyester, this precaution is unnecessary, as this resin does not attack the foam. It even allows you to dispense with an insulating coat of white glue or emulsion paint, if the finish of the core does not require further smoothing. Slight unevennesses can be tolerated, as they fill with resin and as they are only for the inside of the cast anyway.

The good compatibility of epoxy resin and polystyrene foam can also be used to advantage if the core is sealed by a thin coat of epoxy resin. This is allowed to harden before the actual lamination in standard polyester resin and glass fiber. No release agent should be used between epoxy and polyester resin.

The completed fuselage, with its core still inside, is then sanded and smoothed with filler, until it is ready to receive the finishing coat of enamel. Drill into the foam core through an opening, for instance, the cutout for the windshield of the cockpit or for the engine or radio control gear. This can be done with a screwdriver or ripping chisel. Carefully remove some of the foam. Then you can pour solvents (either acetone or methylene chloride) into the small cavity and dissolve the core chemically, which only takes a few minutes.

Dissolving the core

If the core was fitted with an insulating coat of white glue and modelspan paper, this coat can be removed through the hole, too. The solvent makes it soft and elastic, and the release agents prevent it from sticking to the glass fiber. If, however, you did not use release agents, the paper coating cannot be completely removed. If the weight of the molding does not matter, you may leave it in the molding anyway.

This method of work is highly recommended, if you are still experimenting with the design of the fuselage. Once you have found the ideal shape, it is advisable to invest some work into the finish of the prototype molded over the core and then take a female mold from this prototype for molding first-class replicas with an immaculate high-gloss finish. Then, however, you will have to mold the copies in two halves, which have to be assembled later on. Both halves are joined with strips of mat or fabric laminated on the joining line from inside. This is a lasting and, from the technical point of view, unobjectionable solution.

The Female Mold

The working method and the required quality of the cast set special standards for the mold. Such a mold must

- have a resin-proof surface,
- have a sufficiently strong surface not to be deformed or damaged by the pressure resulting from the laminating process and, especially, from the de-aerating of the laminate,
- have the best possible finish, as the finish of the mold will be reflected in the finish of the cast and
- it must allow a trouble-free removal of the cast, preferably without destroying the mold, so that it can be used again and again.

Conditions for a good result

While the three last requirements are more or less self-evident and do not require further explanation, the demand for a resin-proof surface calls for a short explanation, as the mold must be sealed in two different respects. First, the joints of the mold must fit tightly, as we are working with a material that is still liquid for quite a long time in the mold.

Take high-quality Plasticine, which is widely used by artists for sealing the joints of the mold. It seals leaky joints and can be smoothly and easily applied to the mold. It prevents the formation of burrs and flashes, which do not look very nice and may make it difficult or impossible to remove the cast from the mold. As Plasticine is a fairly soft material, it cannot be used for filling wide gaps. They have to be filled with cellulose or polyester filler. Cellulose filler (Plastic Wood, Polyfilla, or similar products) must always be sealed with a two-component lacquer after careful sanding, as this type of filler is porous and will otherwise absorb resin.

Seal joints with Plasticine

This is the second reason for a resin-proof mold. A porous mold will absorb resin, which causes trouble in removing the cast from the mold. But there is also another source of trouble, as porous materials contain a certain amount of air, which expands due to the heat generated by the hardening reaction. The expanding air escapes into the resin just before it is about to set. In most cases, it is then too late to remove the small bubbles in the laminate by rolling.

No porous surfaces!

Which Materials Are Suitable for Mold Making?

The range of suitable mold-making materials is wide. It includes all materials that are by nature hard and dense and have a smooth surface, such as tin (if possible, deep-drawing quality if sheet steel is used, but also sheet brass or aluminum), melamine-coated chipboard, laminated sheet plastic (Formica or Micarta), wood sealed with two-compound lacquer, fiberboards with melamine coating or sealed with lacquer and, most of all, plastics, which must, however, be resistant to styrene, if polyester resin is to be used.

Suitable plastics are glass fiber reinforced polyester or epoxy resin, PU casting compounds, silicone rubber or even latex for castings. There also are low melting masses similar to sealing wax, which may also be used for making a mold.

Plaster and cellulose filler are also frequently used as mold-making materials, but both require a sealing coat of two-component lacquer in order to achieve a resin-proof surface. Normal enamels cannot be used for this purpose, as they are attacked by the styrene contained in the resin or will at least swell, so that the finish of the cast is impaired.

Plain boxlike molds can also be linked with self-releasing Melinex or Mylar film or foil. But take care that the film does not slide or wrinkle while you are working.

Start with a Master

You will rarely succeed in making a female mold without having first made a master. There are a few exceptions to this rule—such as molds for rectangular containers and boxes and fiberboard molds for boats with angled section hulls, as described on pages 65–69, which allow you to build the female mold straight away. Normally, one starts with a master, the finish of which must be as good as the desired finish of the cast.

Seasoned wood   Well-seasoned wood is the most widely used material for making a master. Balsawood is most suitable, being easy to carve and shape, but you can also use European soft wood.

When planning the master, you should remember that the glass fiber cast has to be removed from its mold later on. Complicated masters with protruding edges, pieces or corners or with undercuts may have to be cast in parts, or, at least, the mold will have to consist of several pieces, so that the cast can be removed in one piece later on. Molds that widen in the direction of the opening are useful for easy removal of the cast. It is also most important that all the walls of the mold are stiff enough to prevent them from bowing in. Originally straight surfaces of a mold, which become concave because they are not stiff enough, may spoil the effect. Large wooden molds should, therefore, be reinforced by glueing strips of wood onto the outside. Glass fiber molds should have a sufficient wall thickness, but can also be strengthened by laminating reinforcement to the outside of the mold.

The fillets between two sides of the mold that are meeting vertically or at any other angle are most important. Never allow two sides to meet at a nonfilleted angle. Ensure curved edges. It is essential to disperse any loads that may occur, and it is *the* appropriate method to avoid damage to the material under load. On the other hand, there is also a technical aspect, for glass fibers cannot be brought into an angular shape, because they tend to spring back and take a curved line on their own. If the mold does not have filleted corners, there is quite a risk of damage to the laminate in this zone, because there will be a hollow space between the gelcoat and the laminate or there will be a concentration of resin, making the cast brittle in this region. Sometimes, one does see glass fiber parts with sharp-angled corners and edges, but there is reason to doubt that it is the proper use of the material. Such sharp-angled corners are quite often achieved by means of a trick of the trade. The first layer of mat is butt-jointed in the corner, i.e., there is no overlapping. Then the corner is filleted with polyester filler to provide the curved section required for laminating the following layers properly.

Avoid sharp edges wherever two planes meet

But then you will probably run the risk that the butt joint will get fine cracks under load, as the gelcoat is not elastic enough to take the stress occurring under these unfavorable conditions. In addition to that, such a butt-joined edge has very little impact resistance and will easily burst and break. When molding small parts, it is sometimes attempted to fillet the joint between two planes with an extra thick gelcoat. This is not much use either, as the gelcoat will delaminate under load. As the fairly thick gelcoat is not reinforced with glass fiber in this most critical zone, its impact resistance is nearly zero.

If a multipiece mold is required, the master, too, must consist of several pieces. Where to place the separating lines must be decided from case to case. If you are clever, you will quite often find that two pieces are enough, while a less appropriate position of the dividing lines may necessitate three or four pieces.

If the master model is carved from wood, it should be assembled from several blocks corresponding with the separating parts. These blocks are best joined by means of tightly fitting brass rods and matching sleeves, so that the mold is assembled in the same way as you put a plug into a socket. This technique provides accurate matching joints and avoids gaps, which would be repeated in the mold that is taken from the master model and would also appear on the casting laminated in this mold later on.

The master model must be sanded again and again and receives several coats of paint for a high-gloss finish. Paint the individual segments separately to prevent them from sticking to each other and to make sure that the edges are well sealed, too.

The master model can be assembled after the paint has hardened completely. Place approximately 1/32-in. (1-mm.) thick sheet metal between the separating planes of the master model and allow the metal to protrude all around, like the brim of a hat. The sheet metal is held in place by the same rods that are also keeping the blocks together.

Release agents
Then, the master model and the brim are treated with release agents. There are three different kinds of release agents available, as release wax, release lacquer or release spray. If need be, you may also succeed with a good-quality floor polish or a furniture spray which contains silicone.

The task of the release agent is to prevent the resin used for the female mold from sticking to the surface of the master. This is achieved by a thin wax film, a fine layer of silicone or, if release lacquer is used, a thin plastic film (PVC/PVA), which can be easily washed from the mold or the master with clear water. There are now also Teflon sprays, which produce a thin film of Teflon and are equally effective. Teflon is a self-releasing plastic, the nonstick properties of which are well known to housewives and amateur cooks, who have learned to appreciate pots and pans lined with this material.

The classic release agent: wax
The accepted method of preparing the mold is still with release wax, used as a hard wax dissolved in alcohol. The master model is left to dry for a period specified by the manufacturer and the surface is then rubbed with release wax and polished to a high gloss. To be on the safe side—which is a commendable tactic for the beginner—the polished surfaces are now given an additional thin coat of release lacquer, which must also be

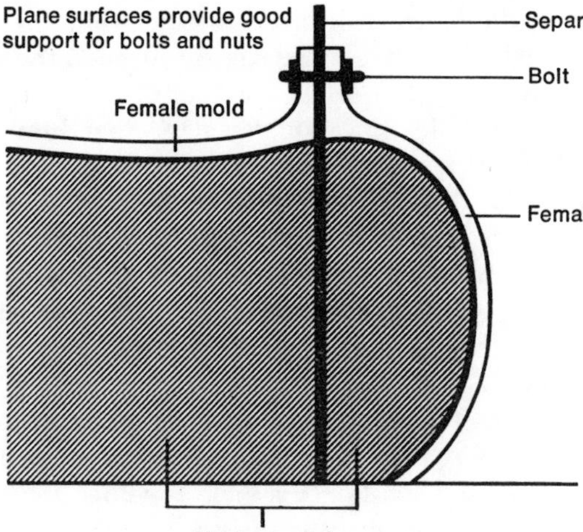

MULTIPIECE MOLD

Plane surfaces provide good
support for bolts and nuts

Separating sheet metal

Bolt

Female mold

Female mold

Master model
(core)

allowed to dry properly. Now, the gelcoat resin is mixed with
hardener and applied to the treated master with a soft brush.
Special care must be taken not to disturb the surface of the master
too much, so as not to damage the film of release lacquer or the
thin coat of wax. After complete hardening of the gelcoat, lightweight glass fiber scrim or twill is laminated onto the gelcoat.
Contrary to glass fiber mat, these woven materials can be quite
easily molded into the rectangular angle between the surface
of the master model and the sheet metal brim. In this case, a
sharp-edged laminate is unavoidable, because otherwise the
mold will not close tightly at this critical joint, where the parts
of the mold will later be bolted together through a glass fiber
brim running all around the separating line of the mold.

This angled joint is then filleted with some polyester filler
mixed with hardener paste to make the following layers of mat
snugly fit the contours of this joint. It is laminated into place
with a fast-curing resin mix and requires constant dabbing with
a brush to keep the mat in close contact with the fillet. Depending on the size of the mold, another two or three layers of
mat are then laminated over the whole surface of the mold.
Very large molds can also be reinforced with woven roving.

When both halves of the mold are completed and cured,
you can drill some holes through the laminated glass fiber brim
of the mold (see sketch above) in order to be able to bolt the

Drill the
metal brim
before
laminating
the mold mold together later on. It is advisable to use a predrilled sheet metal brim and seal the holes with masking tape while the brim is laminated. This makes it unnecessary to drill through the sheet metal and enables you to use a special glass fiber drill.

Having drilled the holes through the glass fiber brim, you may now take the mold apart and remove the master. The separating sheet metal is then no longer required.

The finished molds are now cleaned with warm water and detergents and should be closely inspected. Any bubbles or other faults must be made good with polyester filler. After hardening, they are smoothed with fine, wet abrasive paper. For an extra good finish, you polish the interior of the whole mold with a special polyester polish. Then, the parts of the mold are treated with release agents (again, preferably wax plus release lacquer). Do not forget the flanges! Then bolt the mold together and make sure that all parts are fitting in just the same way they did on the master. A toggle at one of the mounting holes and on the corresponding side of the opposite flange is a great help.

Also, take care that the nuts and bolts are all fitted with the largest possible washers, in order to distribute the pressing forces of the bolts evenly over a large surface of the flange.

Having carefully inspected all the joints of the mold from the inside, you should now seal any fine gaps with Plasticine. The mold is then ready for laminating.

Plaster: Cheap and Easy to Use

Plaster molds are cheap and easy to make. The master is prepared as already described above. Do not forget to apply release agents, but a coat of a good hard wax will do. To achieve an immaculate surface, you should first apply a sort of plaster gelcoat. It consists of a pasty mixture of plaster, which is plasticized by adding some gum arabic or starch. You can also use cellulose filler* instead of plaster and then dispense with gum arabic or starch. Another useful additive for plaster is wallpaper paste, which also plasticizes the mixture.

Make a not too thick mixture and apply it with a brush onto the master model to produce a 1/16-in. to 3/32-in. (1-mm. to 2-mm.) thick gelcoat. Once hardened, this first coat should be wet with a moist brush and then covered with a layer of

* For instance, Plastic Wood or Polyfilla.

fabric (coarse linen, cotton, etc.) soaked in plaster paste. This layer reinforces the mold and helps to prevent cracks. Allow this coat to turn fairly hard and then apply a pasty mixture of plaster, until the mold has the required thickness. You may also put the master with the gelcoat and the reinforcing layer of fabric into a watertight box and fill the box with a not too thick plaster mixture. The separate pieces of split molds are carefully smoothed. Cast the single parts of the mold one after another and allow them to dry for a few days. Any holes or bubbles are filled with some plaster, which is applied with a small spatula. Sand carefully after drying. The well-dried mold must be sealed inside with a good two-component lacquer until the lacquer is no longer absorbed by the plaster and the whole surface has got an evenly glossy finish. The mold is then touched up by final sanding with very fine wet and dry paper and given a final coat of lacquer before release wax and release lacquer are applied on the hardened lacquer. Then, this mold, too, is ready for use.

This method can also be used for molds for large parts. The gelcoat should be about ¼-in. (5-mm.) thick, and instead of the light textile fabric used for small parts, use jute or hessian impregnated with a thin mixture of plaster.

For very small molds, one can make a plaster mold without reinforcing fabric. After the plaster gelcoat is applied and slightly hardened, the master model is put into a molding box, which is then slowly filled with a fairly thin plaster mix, until the separating plane is reached.

After hardening and drying, the mold is treated in the way described and sealed with lacquer. The separating surface is treated with wax, and the second half of the mold is then cast in the same way, until it reaches the waxed separating surface of the first half of the mold. If the molding box is high enough, you can also put the finished half of the mold on the bottom of the box and then cast the second half by pouring the plaster mix on top. Once the plaster has set, the mold can be taken apart in order to treat the second half of the mold in the described way after the removal of the master. After several coats of lacquer and release agents, the plaster mold is then ready for use.

Flexible Molds for Awkward Casts

One can dispense with a multipiece mold, requiring much skill and a high degree of precision, by using flexible resin for

the mold. Due to its flexibility, this material allows even awkward casts with many undercuts to be cast in—and safely removed from—a one-piece mold.

Silicone rubber and PU resins are most suitable for such a mold. See pages 145 and 166–169 for details on how to use these resins. We will now deal only with the technical details of making a mold from these materials.

The question of whether to use silicone rubber or PU resin is primarily decided by the surface structure of the master and, secondly, by the cost of the resin. As silicone rubber is self-releasing, a mold from this material will not require release agents. Silicone rubber molds allow the reproduction of the finest details of surface structures, such as wood grain, unobscured by release agents, which may blur very delicate textures.

Silicone rubber mold

Highly elastic PU casting compounds are equally suitable for more or less level surfaces. However, they require the use of release agents. PU compounds are about one-half the cost of silicone rubber and are therefore preferred whenever the technical requirements allow it. In order to achieve a bubble-free mold surface, a thin first coat (approximately 3/64-in. to 3/32-in. [1-mm. to 2-mm.] thick) should be applied with a soft brush. When working with PU resin, care must be taken that the thin film of release agent on the master is not damaged by the brushing, as this would cause the mold to stick.

PU mold

After the first thin coat has been applied and allowed to set, more resin is mixed and applied slowly. Air bubbles are opened with a toothpick or guided to the edge of the molding box, where they cannot do any harm.

To save resin, the flexible part of the mold should be only thick enough to withstand being peeled off from the cast. Generally, such a flexible mold should have an overall thickness of approximately ⅜ in. (10 mm.). After curing, the mold is cased in plaster, with the master still inside. Before placing the mold into a molding box, make sure that there are no undercuts on the back of the elastic mold that might prevent you from taking the mold from its plaster support. Such undercuts have to be smoothed with resin.

This way, you receive a solid plaster block supporting the elastic mold and allow trouble-free working. After the hardening of the cast, the elastomer mold, together with the cast inside, is lifted from its plaster support. The cast is then carefully taken out of the mold by slowly peeling the elastic skin of the mold from the protruding edges of the molding.

Flexible molds are suitable for laminated as well as poured polyester and epoxy resin casts with or without filler or PU compounds or foams.

When working with polyester resin, it must be remembered that silicone rubber may slightly swell due to the influence of the styrene contained in the resin. Such a slight swelling is not very dramatic, as the styrene absorbed by the silicone rubber is going to revolatilize, after which the mold will return to its original dimensions. The swelling is only a disadvantage if one wants to take a number of casts from one silicone rubber mold; this will then absorb more and more styrene, so that the dimensional precision of the cast will be affected.

Styrene makes silicone rubber swell

But you can also use the swelling of the mold to advantage if you want to make a larger copy of the model.

The Growing Mold

Carbon tetrachloride (caution: vapors are toxic) makes silicone rubber swell to a much greater extent than styrene does. Trichloro-ethane, a cleaning agent that is widely used when working with polyester or PU resin, has nearly the same effect as carbon tetrachloride but is less dangerous. After only a few hours of immersion, a silicone rubber mold will grow to twice its original size. You must accept that the enlarging of the mold caused by the swelling of the silicone rubber will not be absolutely even, as a mold rarely has an even overall thickness. A certain degree of distortion is more or less unavoidable. Like styrene, carbon tetrachloride and trichloro-ethane are absorbed by the mold, which causes the swelling. As both substances are easily volatilizing solvents, the mold will eventually shrink back to its original size.

Large-size Matrices

When reproducing a large and only slightly undercut master, it is advantageous to reinforce silicone rubber or PU matrices with a layer of fabric on the back. Coarse glass fiber twill is most suitable for this purpose. The fabric is laid onto the rear side of the silicone or PU mold before the resin has started to cure. Dab the fabric with a brush to ensure a proper joint between the fabric and the resin of the mold.

Matrices cast from PU compounds are excellent molds for decorative concrete panels having a deep structure. For this purpose, the outside of the rubberlike elastic matrice is frequently reinforced with glass mat, which is laid onto the fresh resin of the cast and slightly sinks into it. Once the PU resin is cured, a further layer of glass mat can be laminated on top with polyester resin containing the required quantity of hardener. This will stiffen the matrice without interfering with its easy removal from the set concrete. By the way, PU matrices are self-releasing, once the concrete has set.

Compound Molds

It may happen that the greater part of a cast is quite flat and easy to remove from the mold, with only a small area that is undercut or intricate and likely to present difficulties in removing the cast. In order to ensure easy removal and to avoid having to make an expensive mold, it is common practice to construct the mold from different materials. You may, for instance, insert loose elastic parts cast from silicone rubber or polyurethane into a simple molding box made from wood, sheet metal or melamine-coated chipboard. These inserts are taken out of this mold, together with the cast, and are then removed from the cast by making use of the elastic properties of the mold-making material, allowing the inserts to be used again and again, as they suffer no damage.

This technique, however, asks for a fairly high degree of accuracy in making the mold. Great care must be taken to achieve a jointless union of the flexible inserts and the surfaces of the main mold. Otherwise, you will have to touch up each casting coming out of the mold.

Colored Gelcoats Save Painting

As already mentioned, a gelcoat is generally applied first when a cast is built up in a mold. The gelcoat becomes the surface of the cast. Gelcoat resins are specially formulated for this purpose. You can buy them as clear, slightly yellowish or colored resins.

No further treatment of the surface required

If you work carefully, the use of a colored gelcoat will allow you to dispense completely with any finishing treatment, such as painting after the cast is taken out of its mold. The only thing you have to do is to remove remains of the release agents from the cast surface. Clear warm water with a dash of a wetting agent (such as a dishwashing detergent) will do. If you want to obtain a high glossy finish of the moldings, there is a special polyester polishing paste.

The amateur is well advised to use only clear resin for the laminate. Clear resin, not tinted by added color paste, makes it much easier to detect air bubbles in the laminate, which can then be removed by rolling.

The use of highly colored laminating resin should be left to experienced professionals, who, due to their long experience can manage to produce a bubble-free laminate straight away. Above all, it is controversial whether a laminate built up with colored resin all through really is advantageous. It is a fact that deep scratches that go through the gelcoat are less obvious if the laminate is built up from tinted resin. But there is no advantage when the damage is going to be mended, since you will have to paint the part in either case after filling the scratch with polyester filler, except if the molding happens to be white, in which case you can use white polyester filler for repairs. In either case, however, it does not matter if the laminate is colored or not.

Colored
laminates are
less suitable
for boats

With boats or other laminates that are constantly exposed to water, an all-through tinted laminate may even be a disadvantage, as you may possibly not notice a damage caused by a collision or underrate its depth. It then may happen that the reinforcing glass fibers are no longer completely embedded in resin, so that water can penetrate into the laminate along the single glass fibers due to capillary action. This impairs the bonding between the glass fibers and the resin and reduces the strength of the laminate.

Clean Tools—Clean Job

Only a few tools are required for making glass fiber reinforced laminates: a brush, a sheepskin roller, a metal-disc roller, several spatulas, some pots for mixing and measuring and a pair of scissors for cutting the glass fiber products (see drawings of equipment on pages 19–20 and 43–44). Like most liquid plastic resins, polyester and epoxy resin are quite sticky liquids, and, as they are unkind enough to harden within a fairly short time to a hard and quite insoluble material, it is most important that the tools be cleaned at once. If you fail to do this, your tools are likely to be fit for nothing but the garbage can.

Sheepskin
roller for
laminating

Tools that have not been properly cleaned will cause bad

results. There are several solvents with a high dissolving power that can be used as cleaning agents. First, you can use acetone, which is rather cheap but also highly flammable. Styrene is another choice. You know this strong-smelling liquid as a constituent of unsaturated polyester resin. Styrene is less flammable than acetone but much more expensive. Methylene chloride is often preferred, as it is nonflammable. This also applies to 1,1,1 triethane chloride, which is a very good solvent for polyester and epoxy resin, but unfortunately this powerful solvent is toxic so that you must ensure good ventilation of your workshop. Finally, as a last resort, you may also use nitro thinners.

All solvents volatilize very easily, so that a substantial quantity gets lost by evaporation if the containers are left uncovered. Because of the smaller surface, tall, narrow containers are more suitable than wide ones, provided they are wide enough to accept the tools. When working with resin, keep the solvent
containers sealed and also cover the pots used for cleaning the tools—with a lid, a piece of wood or sheet plastic. The high specific gravity of methylene chloride can be used to advantage if you pour some water onto this solvent. As water has a lower specific gravity, it will settle on top and form a liquid lid, preventing the solvent from volatilizing and, at the same time, keeping the solvent accessible for immersing your tools for cleaning.

A further hint will make your work much easier. Use three pots of equal size placed side by side instead of only one pot for cleaning your tools. The first pot is used for the precleaning, the second one for the first rinse and the third one for the final rinse. It is quite understandable that the first pot will soon be contaminated with resin, as the dissolving power of the solvents is quickly exhausted by fairly large quantities of resin introduced with each tool. The solvent becomes opaque and begins to produce a deposit, an indication that the solvent is exhausted. Then the first pot is emptied, washed with solvents and filled again to become the third pot in the row, while pots number two and three will move up accordingly.

This brings us to the question of what to do with exhausted
solvents. Do not pour solvents down the drain (risk of explosion). For the same reason you should be careful not to pour solvents into the toilet; otherwise, paying a visit might be the end for a smoker. If you think that funny because you are a nonsmoker, you should still remember that the sewage pipes of many modern buildings are made from plastic tubing. The idea that the aggres-

sive solvents might attack the plastic pipes is not very pleasing, and it may turn out to be a very expensive way of disposing of solvents.

Because of the high risks of explosions, it is not advisable either to burn solvents or to pour the remains into an empty can that is then thrown into the garbage can. In the latter case, you run a double risk: as many garbage cans are made from plastics today, a leaky container or one that cracks when the garbage in the can is compressed by tight packing will allow the solvents to run into the garbage can and dissolve it. Above all, there is a fire hazard and the risk of an explosion if embers or other burning materials are thrown into the garbage can or if the contents of the garbage can are carried to a garbage incineration plant.

There is one way left to get rid of the awkward remains of used solvents. Pour them into flat can lids or containers and let them volatilize. But only do so if the place does not allow the solvents to trickle into the soil and any fire hazard is completely excluded. For someone who is conscious of modern pollution problems, this might not appear as a perfect solution, but this method seems to be the least dangerous one, especially if one bears in mind that the waste from do-it-yourself jobs rarely runs into tons. Craftsmen using bigger quantities had better ask the local authorities for the least dangerous way of doing away with used solvents, in order to avoid pollution problems.

Protect the Floor!

Many do-it-yourselfers and hobbyists work enthusiastically with liquid plastic resins in their home, in a cellar or in their garage. Whoever does so should consider that the floor should be protected with a thick plastic film or foil or brown paper, for it may easily happen that a mix of resin is dropped on the floor or that activated resin trickles on the floor and that this mishap is not noticed. If one works in a garage or cellar, it is advantageous to give the floor a protective coat with a one-component PU sealing (see pages 131–133) and treat it with slip-proof floor wax after the coating has turned hard. You then have a fair chance to chip the hardened resin stains off. People who carelessly fumble with resin and those who have bad luck and detect resin stains even though they tried hard to avoid them still have a chance to remove hardened stains from stone and concrete floors. There is a pasty mixture of very aggressive solvents that

How to remove
resin stains

even manage to dissolve hardened stains of resin. This paste is put onto the resin and then covered with aluminum film or foil. Allow a few hours for it to take effect. In most cases, the resin will then have sufficiently softened so that it can be removed.

Do not rely on this paste if you have soiled your clothes with resin. Do not allow the resin to turn hard, but remove it at once with the solvents mentioned above. The solvent paste will ruin your clothes by producing other permanent stains. An old pair of overalls or expendable clothes and shoes are much better, for the old saying "An ounce of prevention is better than a pound of cure," also applies to working with resin.

A Concise Guide to Laminating

THE MOLD: The cast can be only as good as the mold. Therefore, no pains should be spared where the finish of the mold is concerned.

RELEASE AGENTS: Before laminating a cast, the mold must be treated with release agents. The best method is to use both release wax and release lacquer. Apply the release wax, allow to dry and polish with a soft rag to a high-gloss finish. Make sure that the rag is free from fluff, which might settle on the mold. Finally, apply an even coat of release lacquer with a soft brush and allow to dry.

GELCOAT: Mix gelcoat resin with hardener and paint the mold with a broad, soft gel mix brush, using long, quick strokes. The final thickness should be approximately 1/64 in. (0.4 mm.). Allow to cure according to the manufacturer's specifications.

INSULATING LAYER: Laminate a layer of polyester fiber or light glass fiber scrim onto the gelcoat. Use the minimum amount of the resin mix to impregnate the scrim. Remove bubbles by rolling.

LAMINATE: Build up the laminate with chopped fiber mat and/or glass fiber fabric in the sequence shown in the plan or according to the technical requirements. Cut the mat into handy pieces and fray the edges before adding the hardener. Mix only as much resin as you can use during the specified pot life. Adhere to the mixing proportions.

Apply mat or fabric dry on the preceding layer and impregnate approximately 10 sq. ft. to 15 sq. ft. (1 m. to 1.5 m.) at a time. Use a sheepskin roller soaked with activated resin.

Flatten any air bubbles—indicated by light patches—before the resin begins to gel. If you have an assistant, share the work—one can impregnate the glass fiber while the other wields the roller.

Overlap the joints and take care that the joints in the single layers are staggered.

Always embed woven roving between two layers of mat.

PRESEALING COAT: Once the final layer of the laminate has turned hard, it is given a coat with a mixture of resin and hardener tinted in the desired shade of color by adding 10 percent to 15 percent of pigment. This coat is rolled on and provides the inside finish of the molding. This presealing coat ensures a complete embedding of the glass fibers and improves the resistance of the molding against weathering and water. At the same time, it acts as a first coat of paint, supporting the covering power of the finishing coat of air-drying polyester resin.

The presealing coat provides a thicker coat of resin and still better protection if one layer of polyester fiber scrim (0.065 oz./sq. ft. [20 g./m.²]) is applied at the same time.

FINISHING RESIN: A coat of air-drying polyester resin makes the slightly tacky surface of the presealing coat nontacky. The clear resin can be tinted to any desired shade by adding color paste. If the presealing resin has already been tinted with color paste, 5 percent of color paste will do for tinting the air-drying resin. If clear resin was used for the presealing coat, add 10 to 15 percent of color paste to the air-drying resin. Mind the minimum working temperature (in most cases 64.4° F [18° C]) and do not forget to add hardener. The film of air-drying resin must have gelled within fifteen to thirty minutes after it was applied in order to cure properly. Otherwise, too much styrene evaporates, and the surface is not sufficiently protected against weathering and water.

REMOVAL FROM THE MOLD: Patience is a most reliable assistant for removing a laminate from its mold. The more the hardening proceeds, the more the shrinking of the laminate (approximately 2 percent in each direction) promotes the loosening of the cast. Wherever the laminate separates from the mold by its shrinking, water can be poured into the gap. Moldings with a flange can be loosened from their mold by forcing a piece of stripwood between the edge of the mold and the corresponding flange of the cast and carefully using it as a lever. When casting large parts, it might be helpful to laminate a large wooden block into the cast. Then, screw a strong eye bolt into the block, and suspend the cast by a rope fastened to this eye bolt. The weight of the cast aids its removal from the mold. Always place something soft under the cast in order to prevent it from hitting the floor and being damaged if it suddenly slips out of the mold.

CLEANING CASTING AND MOLD: Remains of release agents are washed away with warm water and a soft sponge, possibly adding a wetting agent, or a mild nonabrasive household detergent or even liquid soap. Rinse with plenty of clear water.

FURTHER TREATMENT OF THE MOLD: Check the mold for faulty surfaces and sections where the filler has come off. Apply new filler after the mold is completely dry again. Smooth repaired sections by sanding. Remove any faults in the resin finish by sanding and touching up the resin. Apply a new coat of wax and PVA or just wax. Now the mold is ready for the next casting. Experience has shown that the removal of casts from the mold becomes easier the more the mold is used; it is then "run-in." Nevertheless, you must never forget to apply release agents before using the mold.

TRIMMING THE MOLDINGS

Quite often, the upper edge of the mold will be slightly higher than was required for the actual height of the cast. This additional height is trimmed off after the laminate is taken out of the mold, but it ensures that the very edge of the cast has the full wall thickness and is free from faults.

Trimming with an oscillating saw or abrasive wheel

The best tool for trimming a molding is a fretsaw (oscillating saw) fitted with a hard-metal-tipped saw blade or with an abrasive wheel. Normal saw blades will get blunt very quickly, as the content of glass in the laminate damages the saw blade very much. The oscillating movement of the saw blade of a fretsaw, as well as the strain produced by the use of an abrasive wheel, are quite unfavorable, as they may lead to delaminating, i.e., splitting of the laminate into single plies. In certain circumstances, the plies will only delaminate in a very small zone but later extend under load. In order to keep such risks as small as possible, special care should be taken that the cutting tools are very sharp and that the laminate is supported or clamped where the cut has to be made. This may be done by clamping the laminate between two supporting lengths of stripwood held together with C-clamps. They could possibly serve as a jig for cutting at the same time and thus make it easier to make a straight cut. At the same time, they stop the laminate from moving while it is being cut.

Casts with a flange generally have a slightly uneven edge. In such a case, the trimming is best done while the cast is still in its mold and the resin is not yet hard, but only gelled to a leatherlike state. The edge can then still be quite easily trimmed with a sharp knife guided along the edge of the mold, being moved like a saw. A sturdy pair of scissors, like those used by tailors, can also be used for trimming the edge of a molding, as long as the laminate is still in the semigelled state. If the face of the laminate has to be smoothed, an oscillating sander, put on with its grinding pad oscillating in the longitudinal direction of the laminate, is more suitable than a rotary grinder (rubber disc), as the sanding pad oscillating in line with the laminate does not produce forces that may split the plies and cause delamination.

Oscillating sander

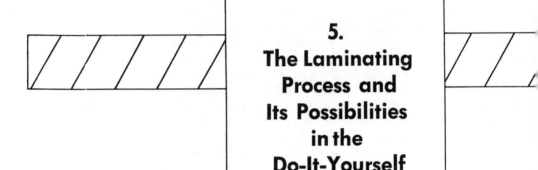

5.
The Laminating Process and Its Possibilities in the Do-It-Yourself Field

Even today, the dream of one's own boat has quite often to be abandoned, as the costs are much higher than the hobby budget. Sometimes, such a dream comes true even though the purse of the boat owner is hollow-cheeked, but all too often, reality falls a long way short of the dream. The small motor cruiser is reduced to the size of a nutshell drifting slowly and clumsily, or the speedboat has become an inflatable contraption, which might be trim but is nothing in comparison with a high-speed rigid boat. This is said without any disrespect to inflatable boats or their owners.

HOME-BUILT PLASTIC BOAT BUILT IN A FEMALE MOLD

Building one's own boat as a do-it-yourself job from plastic resin is an alternative to this dilemma and helps you to kill two birds with one stone: first, you can save money by investing your own work and time, and, in addition, you will get a boat made from a highly resistant material requiring little maintenance or care and looking very attractive—which also explains why this material is used more and more by professional boat builders, too.

The amount of money you can save by building your own boat, compared with a commercial boat of the same size, depends on the size of the boat. With a small (approximately 13-ft. [4-m.] long) open boat for four people, you may save approximately 35 percent to 50 percent by laminating the glass fiber shells for the hull and deck and by finishing the assembled boat yourself. If you start from a basic price of about $1,800 (£700) for a commercial boat of that size, the saving will cover a large part of the costs for an outboard engine.

Saving of expenses

Many people might say that such a saving must of course be paid for with an inferior quality boat. But this is by no means a valid objection if the amateur has average skill. The method is exactly the same as used by professional boat builders: you work in a female mold, which guarantees an immaculate finish. The special method of working is the hand-lay-up process described in detail on pages 41–42 and in a brief outline on page 47. Any skilled amateur will be familiar with this technique after only a few small-scale trial jobs, and he can achieve perfect results, especially as there are many detailed instructions dealing specifically with boat building. In this book, therefore, we shall confine ourselves to describing the various possibilities for the benefit of amateurs planning to build their own boat and to showing ways of avoiding possible risks.

No inferior quality

Plans, Mold Making and Hired Molds

Inexperienced amateurs should, in any case, refrain from making their own designs and, instead, copy well-tried designs, which may be adapted to one's special requirements. This is a sure way to avoid failure. There are plans for all kinds of boats, ranging from a simple fishing boat to a racing catamaran or 31-ft. (10-m.) motor yacht. All these plans are specially designed for polyester and glass fiber construction, and they include all information required for building such a boat, including the building of the mold, the structure of the laminate, necessary reinforcements and the final completion of the interior or rigging plans for sailing boats. Specially prepared for do-it-yourself boat builders, these plans avoid design features that cannot be

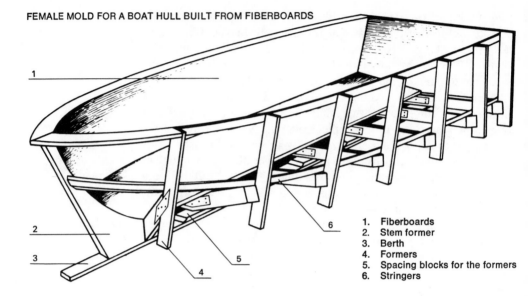

1. Fiberboards
2. Stem former
3. Berth
4. Formers
5. Spacing blocks for the formers
6. Stringers

Home-built mold

realized by an amateur. In order to achieve immaculate laminates, it is indispensable to work in a female mold, as already mentioned above. You can either build such a mold yourself or rent one. An ideal way of building one's own mold is the so-called Voss method, based on a wooden skeleton of formers joined with stringers and lined with fiberboards.

This tried and approved method only requires average skill. The skeleton of the mold is built up from roof battens and consists of a number of angled formers, joined with plywood stringers nailed against the formers from inside to stiffen the skeleton. Fiberboards or resin-impregnated shuttering panels cut to the desired shape by using cardboard templates are then nailed down onto the stringers. The joints between the single panels are filled with cellulose, or polyester, filler that is also used for fillets, which not only create graceful lines but also provide the rounded edges indispensable for a good laminate.

The inner surfaces of the mold must be carefully sanded to a smooth finish. Fiberboards require two coats of a two-component PU lacquer (DD lacquer). If you use the more expensive resin-coated shuttering panels or formwork, the application of filler and lacquer only applies to the filleted joints between the single panels.

All nails must be carefully countersunk. The resulting cavities have to be filled with polyester filler. It is also most important for the mold to be erected on an entirely level plane to avoid warping, which would inevitably affect the running characteristics of the boat.

If your workshop does not have an even and level floor, each and every former must be adjusted with a spirit level, unless you make up your mind to erect a level platform on the uneven floor.

Building a female mold from fiberboards and a supporting skeleton of battens and stringers does not greatly tax a handyman's skill, as the woodwork is fairly simple. Precision, however, is a must: any inaccuracy or carelessness in finishing the mold will show up in the boat and cause a lot of work—always providing such faults can be corrected at all.

As building one's own mold requires quite a lot of time, and as the mold, once used, will be of little value or even take

DIFFERENT WAYS OF STIFFENING A GLASS FIBER MOLDING BY A CLEVER DESIGN AND THE CORRECT SHAPE OF THE EDGES.

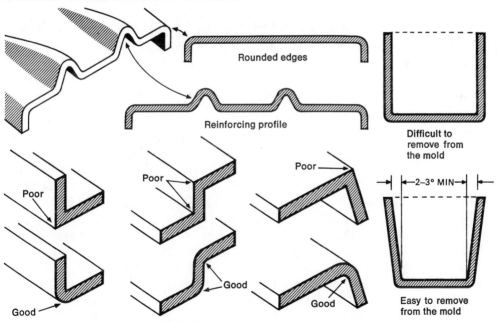

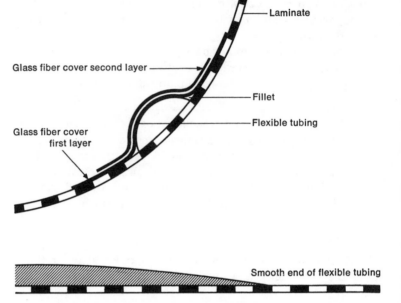

Longitudinal stiffening with flexible tubing (see pages 71–74 for description).

Rented mold

up space required for the boat, many do-it-yourself boat builders prefer renting a mold. Several aspects are in favor of this alternative: it saves time, as you need not build the mold yourself; it saves the amateur, who is not a dedicated woodworker, the use of hammer and saw and—the most important point— guarantees a really perfect finish, as molds for rent are made from glass fiber reinforced polyester and have a perfectly smooth finish.

A rented mold has another technical advantage: it enables you to build a curved-section hull, which is superior, from the hydrodynamic point of view, to the sharp-angled designs built in a fiberboard mold, which would make the building of a curved-section hull too complicated. There also are a great variety of types and sizes of boats that can be built in a rented mold.

Renting costs and time

You can work at home or in a workshop of the renting company, which will also supply you with all the materials you need. The rent for a mold is always quoted per week and may be, for instance, about $130 (£55) per week for, say, a 16-ft. 4-in. (4.3-m.) long open runabout. An extra day will cost about $20 (£9).

If one works on the site of the renting company, one can save the transportation costs for the material and the return of the mold, for instead of transporting the mold twice, you only have to bring your finished boat shells back home. A further advantage is the easy availability of all materials required and the chance to get some hints from an expert now and then.

With regard to time, seven days should be plenty of time to laminate two shells (hull and deck) without running into trouble. If you are a handy person and have a useful assistant, you may even be able to make two complete sets of hull and deck in one week. This prospect makes it worth considering whether you cannot find a friend who might be interested to join in. Such teamwork brings you a double advantage: you save renting costs and also buy the materials at a cheaper price due to a bulk reduction.

Build your boat in one week

It must also be mentioned that one has to pay a deposit when the mold is rented. It is refunded in full when the mold is returned in intact condition.

Taking a Mold from Another Boat

If you are well acquainted with polyester and glass mat, you have still a third possibility. You can take a mold from another wooden or plastic boat. There is, however, the prerequisite condition that the boat owner is willing to let you have his boat for this purpose. It is self-evident that special care must be taken when treating the boat with release agents. The best material for taking a mold from a boat is polyester and glass mat or woven glass fiber. If the boat is not too big, you may also use jute-reinforced plaster, which may possibly cause less concern to the boat owner than taking the mold with polyester and glass fiber.

Before starting work, make up your mind if a two-piece mold is required. Splitting the mold later on in order to be able to remove it will damage the boat, however careful you are. If a two-piece mold is absolutely necessary, you will have to cut a closely fitting template from thick plywood, which must not be warped, and put it on the hull of the boat where the separating line is going to be. Then laminate against this template when making the mold and allow the laminate to run up the template to form a flange, required for bolting the two halves of the mold together later on. The second half of the mold is made in the

Subsequent splitting of the mold may damage the master boat

same way, but do not forget to shift the template by the amount of its own thickness toward the side the first part of the mold was taken from. This is necessary to ensure a precise fit when the two halves are assembled.

Anyone thinking about taking a mold from an existing boat should deliberate on two principal questions:

- Is the design suitable for glass fiber?

- Could taking a mold from this boat possibly infringe upon the rights of the designer?

The first question is most important if the master is not built from glass fiber but from wood or even metal. When copying a wooden or metal boat, at least a filleting of any angled corners and rounding off of sharp edges will be necessary, but this may have some influence on the running characteristics of the boat. In such a case, the advice of an expert is desirable. Even inside the boat, the use of glass fiber and polyester may have its consequences, as the standard stiffening of a wooden or metal hull with bulkheads cannot be simply taken over for a plastic boat.

The thickness and the sequence of the different plies of the laminate will also play an important role in such a case. You may obtain reliable advice from the supplier of the materials. You may also possibly send a three-dimensional view or some photographs of the boat you intend to copy to such a firm to make sure whether this is a suitable prototype for a glass fiber copy.

Legal problems Copying a design also involves possible legal consequences. Generally, there are no objections against a noncommercial copy, but special care is recommended with specially designed prototypes and racing boats. In such cases, the consent of the designer or the manufacturer is needed, for the OK of the boat owner does not mean very much. In case of a lawsuit, it would only make him a co-defendant.

In this respect, a nonplastic prototype presents fewer problems, as adapting it for glass fiber will generally call for alterations of the design, so that one can no longer call it a copy. If you want to avoid any trouble right from the start, you had better build a trial boat from a kit either in a fiberboard mold or in a rented polyester mold.

To design one's boat from scratch requires a considerable technical knowledge of hydrodynamics and the construction of glass fiber parts, and nonexperts would be well advised to refrain from such experiments.

We should like to give a few hints on how to build up the laminate when adapting a nonplastic motor boat for glass fiber.

LENGTH OF BOAT	STRUCTURE OF LAMINATE
approx. 13 ft. (4 m.)	G–M 300–4x M 450–M 300–LT
approx. 17 ft. (5 m.)	G–2 x M 450–R 670–M 450–LT or G–M 300–M 600–M 450–R 670–M 300–LT
approx. 24 ft. (7 m.)	bottom: G–M 300–M 450–R 670–M 600–R 670–M 300–LT
	sides: G–M 300–M 600–M 450–R 670–M 300–LT

G = gelcoat, M 300 = glass fiber mat 300 g./m.^{2} = 1 oz./sq. ft. M 450 = glass mat 450 g./m.^{2} = 1.5 oz./sq. ft. (standard mat), M 600 = glass mat 600 g./m.^{2} = 2.0 oz./sq. ft., R 670 = woven roving 670 g./m.^{2} = 2.2 oz./sq. ft. with uniform tensile strength in both directions, LT = air-drying polyester finishing resin. Surfaces are improved by applying one layer of polyester fiber scrim after the gelcoat and another one before the finishing coat of air-drying resin is brushed or rolled on.

Some Limits on Construction

For transforming a design based on conventional materials into one suitable for polyester and glass fiber, some construction hints are quite useful. They might also be of help in planning other work.

Due to the high elasticity of glass fiber, oblong moldings, such as boat hulls, require some stiffening. Some of it is contributed by the deck, which is joined to the hull with strips of glass mat laminated over the joint. With larger boats, this way of stiffening will not be sufficient, so that additional stiffening

Oblong moldings require stiffening

must be considered. Firmly mounted objects, such as bulkheads bolted or glued into place in metal or wooden boats, cannot be recommended for glass fiber boats. They reduce the elasticity of the laminate in certain regions of the hull. Under load, these parts would not be resilient enough. This causes stress concentrations in certain zones of the glass fiber shell, which may give rise to cracks or delamination.

Longitudinal stiffening is most appropriate for glass fiber, and it is achieved either by a clever design stiffened with corrugated or V-shaped profiles in the bottom of the hull, or even the sidewalls, or by stiffening profiles to the inside of the hull.

Reinforcing profiles with flexible tubing

Such stiffeners are easy to make. Cut a length of corrugated flexible tubing as used for air ducts—cut it lengthwise and fix the two halves to the hull with some polyester filler. This will form a stiffening profile with a semicircular section. The flexibility of this tubing allows you to achieve a smooth line, even if the contours of the sidewalls are curved (see sketch on pages 67–68).

The tubing is now coated with two layers of mat and resin. The first one should be laminated over the tubing and form a flange of a hand's breadth on each side. The next layer should be a bit shorter, resulting in a staggered joint, which is indispensable wherever the laminate varies in thickness, i.e., even where the thick laminate of the bottom of a boat joins the less thick sidewall of the hull. This also helps to avoid stress concentrations.

The flexible tubing is completely coated with glass fiber. Both ends of the tubing have to be carefully cut to shape in order to achieve a graceful joint with the sidewall. Use polyester filler for filleting.

Plastic foam profiles

Rigid plastic foam can also be used (e.g., high box-type profiles for stiffening the bottom of the boat). Here again, the stiffeners are coated with glass fiber and are glued into place.

Wherever loads have to be taken, it is necessary to disperse them over the largest possible area. This can be achieved, for instance, by laminating wooden blocks or sheet metal plates into or under the laminate. As always, overlaps of glass fiber mat should be staggered.

Mounting of cross members and shock absorption

If cross bulkheads are indispensable for any reason, care should be taken to provide a shock-absorbent resilient joint using a strip of plastic foam between the glass fiber-coated bulkhead and the sidewalls of the hull. Cut the foam padding

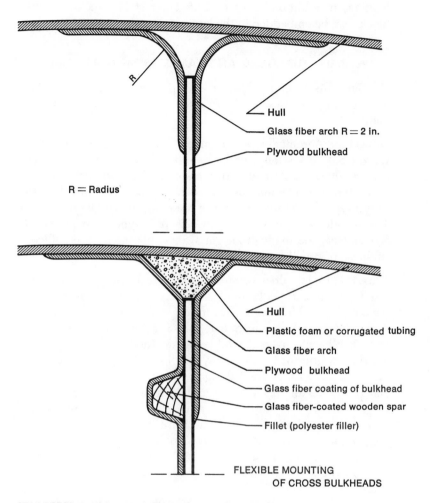

R = Radius

Hull
Glass fiber arch R = 2 in.
Plywood bulkhead

Hull
Plastic foam or corrugated tubing
Glass fiber arch
Plywood bulkhead
Glass fiber coating of bulkhead
Glass fiber-coated wooden spar
Fillet (polyester filler)

FLEXIBLE MOUNTING
OF CROSS BULKHEADS

STAGGERED RUNNING OUT OF REINFORCING LAMINATE

} Normal laminate

into a fillet with a fairly large radius to achieve even dispersal of stress on the hull on both sides of the bulkhead.

If possible, large glued joints and graduated laminated joints with a curved load-distributing flange should be given precedence over punctiform bolted joints, which are typical for wooden or metal structures. If technical aspects nevertheless call for bolting a part in place, use large washers for dispersing the load or, in addition to that, drilled flat steel bars or wooden blocks can be embedded in the laminate.

MODELS, FURNITURE AND LAMPS FROM GLASS FIBER

The outstanding properties of resin-bonded glass fiber allow a wide range of uses. Here are some suggestions for the handyman.

MODELING: Aeromodelers and model boat fans can mold lightweight but very strong fuselages and hulls from glass fiber. The best results are obtained with female molds. Start with a master (prototype), which may be carved from wood and must be carefully prepared before painting or spraying to a high-gloss finish. The female mold is taken from this prototype (see pages 44–56) and can be made from glass fiber, plaster or even silicone rubber (for smallish parts only).

FURNITURE: Glass fiber furniture has become very popular in the last few years. It looks most attractive, but unfortunately you have to pay a fairly high price if you want to buy it in the shop. The price is high, because first, such furniture is only produced on a small scale by the hand-lay-up method, and second, there are the royalties of the designer, for plastic furniture depends on first-class designs. A good design is required to exploit the special properties of the material and to turn out furniture that is elegant, as well as sturdy and serviceable. At the same time, a clever design gives glass fiber furniture its special touch, resulting from its graceful and somehow "organic" lines. From the technical point of view, amateurs will no doubt be able to make their own good-quality glass fiber furniture. The question of design is the main problem.

Styling is important

Anyone thinking of making his own furniture from glass fiber—which is ideal for a garden, as it is highly resistant to weathering—should have some talent as a designer. Studying commercial glass fiber furniture will be a great help. Another important aspect is the proper shaping of the seat and the back rest of chairs and other seating accommodations that must conform to the anatomy of the human body. Above all, your design should allow for easy removal of the cast from its mold. The prototype (master) can be made from wood or plaster. A provisional mold can be taken from the prototype by using plaster and jute (see section "Plaster: Cheap and Easy to Use," pages 52–53). When laminating the first chair or table, you should refrain from sealing the rear side of the laminate with air-drying polyester resin unless you have previously established that the laminate is strong enough to withstand any stresses likely to occur in normal use. If it turns out that the laminate is not strong enough, you may easily apply another layer of mat or fabric or even several layers. If the molding turns out to be strong enough, you may now apply the presealing coat of resin with a layer of surfacing mat embedded in it and finally give the laminate its finishing coat of air-drying resin. If you intend to make several copies of this first model molded in a plaster-jute mold, you should now touch this prototype up by sanding and filling it wherever the finish is not quite satisfactory. Finally, give it a finishing coat of high-gloss DD lacquer, and, if possible, polish it with polyester polishing paste. Then, take a glass fiber female mold from this prototype model.

Test strength in use

LAMPS: Homemade glass fiber lamps are less difficult to design and less complicated and time-consuming to make. As glass fiber is completely waterproof, lampshades and lanterns are a nice subject for do-it-yourself enthusiasts and keen gardeners. There are several fairly uncomplicated methods of building such a lamp. Spherical lamps are laminated over a balloon, the rubber bladder of a football or even an inflatable plastic ball. That means that you work over a male mold. The first layer should be applied with a fast-curing resin mix in case the mold is damaged by the resin. For the first coat, you may use fast-hardening glass fiber paste, which can be easily brushed on. Then, apply normal glass fiber mat with standard resin, which can be tinted to light transparent shades by adding a little color paste to the resin.

Quite interesting effects can be obtained by laminating roving strands onto this layer, which produce ornamental patterns.

Cylindrical lamps are molded over a length of PVC tubing cut down its entire length for easy removal of the lampshade after it has been laminated on the outer surface. Plywood discs are used to keep the slotted mold in its original shape. The longitudinal seam is sealed with masking tape from the outside. A few lengths of masking tape (Sellotape) are wound around both ends of the tube, where the wooden discs have been placed. Some release wax is applied for easy removal of the finished cylinder, which is laminated to the desired thickness. Once the laminate has turned hard, we can screw a large wood screw straight into the center of the plywood discs, which will serve as a handle to pull the discs out of the tube.

Now, take a spatula and force it between the tube and the

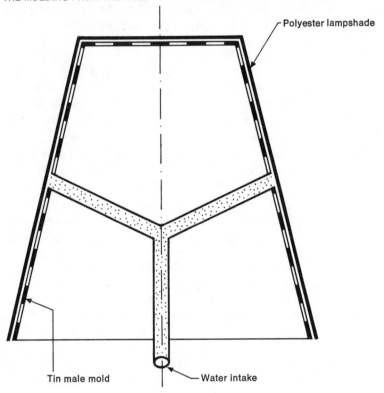

Polyester lampshade

Tin male mold — Water intake

laminate where the tube is split and press the tube inwards. Thus, the diameter of the tube is reduced, and the glass fiber cylinder that was laminated over this tube can be removed. Without this dodge, you would not be able to remove a cylindrical cast from its male mold without destroying either the cast or the mold, as the cast would shrink tightly onto the core.

Such a glass fiber cylinder can then be fitted with a roof in the shape of a flat cone, which can be laminated on a simple cardboard mold covered with self-adhesive plastic film or foil, if necessary. The bottom of the lamp can be made in a wooden mold, but you can also use the lower part of a can cut down to a height of 1 in. (2.5 cm.). Smooth some Plasticine into the joint. The disc to make the foot of the lamp is laminated on a flat sheet of plywood or chipboard covered with Melinex or Mylar film or foil. Remove any air bubbles in the laminate by rolling it carefully with a metal disc roller, then apply another sheet of film. Take care that no air is entrapped when the laminate is covered with this sheet. Once the resin is cured, the self-releasing sheet can be easily removed.

The structure of the glass mat makes the lampshades very attractive—reminiscent of Japanese or handmade paper. It is important not to add too much pigment in order not to reduce the translucency. Above all, ventilation holes are a must, to allow the heat produced by the bulb to escape. Lampshades wound from strands of roving are quite interesting and attractive. A slightly conical shape is to be preferred to a strictly cylindrical one to facilitate the removal from the mold.

CREATIVE WORK WITH GLASS AND RESIN

Glass fiber has much to offer to anyone with an artistic hand. The wall relief illustrated is an example. The master is made from wood or plaster and sealed with two-component lacquer before a female mold is made from silicone rubber or glass fiber. The cast is then made from the female mold. You can also make a master from polystyrene foam, which can be easily shaped. Give it a coating of PU casting compound* or epoxy resin and then reinforce it with plaster or glass mat and resin from the wrong side. Then, you can dissolve the foam with a solvent, treat the mold with release agents and laminate replicas of the master with resin and glass mat. Very delicate structures and fine ribs or other contours too fine for glass fiber can be filled with a mixture of polyester resin and 50 percent quartz powder plus 2- to 3 percent of hardener (based on the quantity of resin) after the gelcoat has been applied and allowed to cure properly. Add as much of this mixture as is required to achieve a fairly even surface, which is then reinforced with glass mat laminates.

* See page 145.

6.
Resin Coatings for Repairs, Work Pieces and Brickwork

The second important way of using polyester or epoxy resin in combination with glass fibers is the coating of surfaces. The method of working is very similar to the laminating process described before, but there is no gelcoat, as you do not work in a female mold. In coating work, the surface is formed by the final coat, as in painting. This is the reason why special attention should be paid to the presealing coat, the embedding of polyester fiber scrim into the last coat of resin and the final coat of air-drying polyester resin (appearance and resistance to chemicals and weathering).

Presealing essential

There is another distinction that is even more important. While easy removal of the coat is one of the main principles when laminating in a female mold, and release agents are used to ensure this, just the opposite applies in coating work, which aims at the best possible adhesion. Mechanical pretreatment of the surface is desirable, to remove loose particles and dust by

Mechanical aids for good adherence

sanding or brushing with a wire brush or with chemicals (e.g., hydrochloric acid for concrete) in order to roughen the surfaces for a good anchorage of the coating on a rough underground. This is recommended for both polyester and epoxy resin.

Epoxy resin adheres very well and rarely requires any aid to good bonding other than a good surface. Polyester adheres badly to wood, concrete, metal and most other materials and requires a bonding agent.

When coating metal, for instance, apply a thin coat 1/48-in. (0.5-mm.) thick of polyester filler (see page 97) as a primer. Due to its high mineral content, polyester filler does not shrink very much. It also contains special additives that improve its bonding characteristics. This intermediate bonding coat of polyester filler serves as a primer and is smoothed by lightly sanding immediately after hardening, before the glass mat is laminated over it with an activated polyester resin. This leads to a chemical interlacing of the fairly fresh coat of polyester filler and the resin.

A second possible method of producing an intermediate bonding coat, mainly used for wood and concrete and other not entirely impervious materials, is based on the excellent bonding properties of PU resins. This works with virtually any material **PU** and takes advantage of the possibility that incompletely hardened **primers** PU resin reacts chemically with the fresh polyester resin. One-component PU resins* have proved to be ideal primers. Owing to their solvent content, they are liquid enough to penetrate into the micropores of the surface and to provide perfect adhesion. Exceptionally difficult surfaces require further thinning of the primer and the application of several coats. It is essential to adhere to the specified intervals between the application of the individual coats of primer as well as the first layer of the laminate. If the time allowed for drying is exceeded, the PU primer has reached a state that prevents a chemical interlacing of the primer with the polyester resin—no interlacing links are left in completely hardened primer. Polyester resin will not adhere well to completely hardened PU resins which explains the working principle of the primer.

Coating with resin-bonded glass fiber is generally carried out to improve the resistance of a material to mechanical or chemical attack. Other aspects include improving the quality

* See page 134 for details.

of the surface and waterproofing or improving its mechanical strength. Some examples of coating jobs will also be of interest to the amateur. Polyester and glass mat are used for mending rusted-through car bodies and other sheet metal parts (gutters and rain pipes). A glass fiber bandage will seal leaky tubings, and polyester and glass fiber will seal cracked concrete surfaces (terraces and old swimming pools); they will also seal and touch up old wooden boats, can be used for repairing leaky flat roofs covered with old tar board and for building swimming pools and ornamental garden pools as a do-it-yourself job.

HOW TO REPAIR RUST HOLES

Many car owners are rather shocked when they detect the first brown pimples in the paint of their beloved vehicle. If they inspect the treacherous spots more closely and slightly press the damaged surface, it will often react with a rotten rustling sound, and then there will be a noticeable hole in the car body where there had been a few spots.

The most vulnerable parts of the car are generally the sills under the doors and the lower parts of the doors if the drainage holes are blocked, the mud protectors in the region around the headlight or headlamp casings and around the joints between the mud fenders and the door, or dead corners on the trunk or boot, where water may easily gather.

With the aid of polyester resin, glass mat and polyester filler, you can banish the ravages of damp and time for ever. Such repairs are by no means provisional solutions nor a botched-up job, but are regarded as a first-class repair job by professional body builders, if—and this is an important if—high-quality resins and fillers are used (as offered in special repair kits developed for this purpose).

The first step must always be to remove the crumbling corroded metal, until you end up at fairly intact material, solid enough to form a basis for the laminate. The whole area is then rubbed with 120-grade wet and dry sandpaper, until there is an approximately 1¼-in. to 2-in. (3-m. to 5-cm.) wide rim of shiny bare metal around the hole. This zone of roughened bare metal around the hole is now coated with an approximately 1/64-in. (0.3–0.5 mm.) thick coat of activated polyester filler. After a few minutes, this coat can be lightly sanded to smooth it for the following layer of resin-impregnated glass fiber mat. This glass fiber patch is cut to the required shape with a pair of scissors, allowing about a 2-in. (5-cm.) margin all around to cover the still intact metal. The edges of this patch are frayed in order to blend smoothly with the car body.

The prepared mat is then placed on an even support covered with polythene or, better still, with Melinex or Mylar film or foil, on which it is impregnated with activated polyester resin by dabbing with a brush soaked with resin.

As soon as the whole piece of mat looks glassy and transparent, pick it up, together with the film or foil, and glue it onto the prepared area as if you were putting a poster against a wall. Then carefully pull off the film, and dab the mat with your brush to remove air bubbles and to make it closely fit the car body.

With most types of car body repair resins, the patch will have hardened after three to four minutes and can immediately be sanded to remove any protruding glass fibers. Apply two coats of polyester filler and carefully sand each coat to achieve a good finish. The surface is then ready for painting. You can either use paint from an aerosol spray can or apply it with a normal spray gun and have it hardened in a heated oven. Good-quality fillers withstand a temperature of 194° F (90° C), so that stove enameling will be no problem at all.

The patch will be hard after a few minutes

The Limits of Polyester Repairs

Cosmetic repairs on your car not only improve its appearance and mean a sort of visual rejuvenation, but they can also prevent further corrosion in the future, as any surfaces coated with polyester resin and glass mat are absolutely weatherproof. Above all, such a coating prevents water and other corrosive media, like salt, from getting to any affected surfaces and continuing their destructive work.

With all repairs, you must never forget the safety of your car. Therefore, the car safety authorities do not allow any glass fiber repairs on load-bearing structure, such as parts of the chassis. The main reason is that it is not possible to be absolutely sure of the strength of such a repair, as this largely depends on how well the repair is carried out. The reservations on repairs with plastic resin on structural parts must, however, not be regarded as a vote of no confidence against this material, for the technicians are also very critical of traditional metal repairs. For some load-bearing parts, for instance, welding is allowed only if the seams run in a longitudinal direction.

In cases of doubt, it is recommended that one ask a special-

No glass fiber and polyester repairs on load-bearing structural parts

ist whether a car repair with polyester resin is advisable. If the answer is no, you must accept it. Going through with the repair nonetheless and camouflaging it with a generous coat of black anticorrosion spray is inexcusable.

NEW BOATS FOR OLD

Even the owner of a wooden boat can use polyester and glass fiber as a reliable repair material that will make his floating vehicle seaworthy again and save it from an inglorious end on a wrecking site. A glass fiber skin also restores the good looks of his boat. Most boat owners are astonished by the fact that their boat now lies higher in the water after the coating with glass fiber and polyester than before, even though such a coating means additional weight. This proves how much water had been carried in the wooden planks. Another by-product of such a face lift is the greater speed of the boat. This is partly due to the reduced draft or draught and lower weight (no more water carried in the wood) and partly the result of the better finish of the boat, if you took the pains of sanding it properly and smoothing it with filler. But this means hours and hours of carefully planned work.

The boat lies higher in the water— despite the additional weight

Here again, the preparations are most important for the success of the coating job. The prerequisite condition is that you must above all allow the boat to dry out thoroughly in order to allow the coating to adhere firmly to the wood. Removal of any old paint serves the same purpose. It is best done with a powerful electric drill fitted with a rubber disc and the roughest grade of carborundum paper. You may, of course, burn the old coats off and then sand the wood, but this method is not so safe, as quite often remains of paint will be left in the grain of the wood and thus impair the bonding of the bonded glass fiber.

Once the boat has dried out well, the joints between the single planks will gape widely. The caulking is then removed with a scraper, and matching lengths of plywood are glued into the gaps with waterproof glue and are finally planed flush with the hull. Smaller gaps are filled with polyester filler and smoothed by sanding. Any fittings or parts that might have to be removed for repairs or maintenance work have to be taken off before you start coating your boat and remounted after the job is completed.

All metal parts that are also coated with glass fiber have to be sanded thoroughly and must receive a thin coat of filler.

Once the complete hull is prepared for coating and all rectangular sharp edges and corners are rounded off by sanding, planing or filleting, as is appropriate for the coating material, you must not forget the most important step: the application of a primer.

PU primer solves any adhesion problems

For years and years, the coating of wooden boats with glass fiber was regarded as a somewhat dubious procedure, as the glass fiber coating sometimes detached itself from the wood. The reason for this failure was poor adhesion between the wood and the laminate, because the hardening reaction was partly impaired by certain substances contained in the wood, because the wood absorbed too much styrene and, finally, because the thick syruplike resin could not penetrate deeply enough into the grain of the wood. Homemade primers made from polyester resin, acetone, styrene, cobalt accelerator and hardener were used to anchor the first coat of resin deeply into the pores of the wood and to insulate the inhibiting substances contained in the wood.

Even though this measure led to quite good results, it could not be regarded as fully satisfactory. Today, this problem is solved. Very fluid and, therefore, deeply penetrating one-component PU resins are used as primers and provide an excellent "anchorage' 'in the wood and ensure reliable and lasting bonding of the glass fiber coating to the wood.

The primer is brushed or rolled on in stages. Always apply an even coat on approximately 50 sq. ft. to 60 sq. ft. (5m.²). Allow the primer to dry for at least half an hour before laminating, but not longer than three hours. Apply the activated resin with a roller before you start to apply the single layer of mat, which should have been torn to size. Impregnate the mat with some more resin, and roll it carefully, to remove air bubbles. The boat is coated in strips, with each strip of mat overlapping the preceding one. Depending on the size and condition of the boat, one to three layers of mat are applied. The last one is lightly sanded and receives a first coat of polyester filler, which is sanded again before any remaining irregularities of the finish are smoothed with more filler. The surface is sanded again before the two vital last coats are applied: first, a sealing coat consisting of a mixture of activated resin and color paste, and second, the finishing coat of air-drying polyester resin. The working time for the laminating process is about 10 min./sq. ft. (or 2 hr./m.^{2}).

It is most important that your boat, or any other wooden object, receive a glass fiber coating on only one side to allow the wood to breathe. Otherwise, the wood may be destroyed by dry rot.

In order to ensure proper adhesion of the coating at the upper edge of the hull, you may laminate the coating over this edge and down one or two inches on the inner side, if the shape of the hull allows this. In any case, the edges must be slightly rounded to make the laminate fit tightly.

Apply a coating on one side only when fiberglassing wood

It is also a good idea to secure the laminate by screwing down the rail with long wooden screws going through the laminate into the wood of the hull. In this case, the sheer rail is removed before the hull is coated with glass fiber and remounted when the coating is completed.

The same method can also be used for fiberglassing other wooden structures, such as garage doors, to make them weatherproof and more resistant.

If You Have a Steel Boat

Even steel boats are sometimes coated with glass fiber to reduce corrosion. In such a case, however, use of epoxy resin is often essential, as polyester resin does not stick sufficiently well to the steel. With motorboats, in particular, the strain caused by vibrations is extremely great, so that only epoxy resin can be used. For coating the steel hull of a sailing yacht, you may possibly be able to achieve a lasting coating with polyester resin by using a PU resin as a primer.

In case of doubt, you should ask your supplier of glass fiber and resin for special advice as to whether a coating with polyester resin can be recommended or not.

COATING TERRACES AND FLAT ROOFS

Cracked and leaky terraces can be restored by a plastic coating. If there are only small cracks or if the concrete is only porous, a simple sealing of the surfaces with a one-component PU resin will do. For a more attractive finish, you can apply a colored coat after the first coat of clear sealing agent or apply a final coat with two-component PU resin or plastic chips embedded in a solution of clear plastic resin.

If cracks develop during setting, they have to be filled with polyester filler. Quite often, such cracks expand only a little or even stop expanding after some time, so that they can be satisfactorily stopped with this relatively rigid filler.

Constantly moving cracks, however, represent more of a problem. In such a case you must use a filling compound that permanently remains elastic, based on PU resin or silicone rubber. In both cases, the crack must be widened and deepened to a depth of approximately ½ in. (13 mm.) to make sure that enough material can be put in to absorb the movements of the concrete. A careful application of the right primer for the filling compound is most important to ensure that the filling adheres firmly to the sides of the crack.

Expanding cracks in a surface and cracks frequently occurring where horizontal concrete surfaces meet brickwork can be efficiently cured with a coating of polyester and glass fiber.

As always, good adhesion is vital. Surplus cement slurry must be removed with hydrochloric acid. The best method for doing so is using a plastic watering can with sprinkler nozzle and a long-handled hard scrubbing brush. Having dispersed the acid on the concrete and having allowed it to react, the surface must be washed with plenty of clear water again and again to remove any remains of the acid. Allow to dry thoroughly (several days of sunshine). Then fill the cracks with polyester filler and sand them until flush. After application of a primer (e.g., G 4) the surface is ready for treatment with polyester and glass fiber. **[Remove excess cement slurry with acid]**

Wherever the surface of the terrace meets a wall, you must make a fillet with a radius of approximately 2 in. (5 cm.). Extend the glass fiber up the surrounding walls to a height of 4 in. to 8 in. (10 cm. to 20 cm.), where it should end in a gap cut into the walls. This groove is cut into the plaster of the walls with an abrasive wheel and should be about ½-in. (1.2-cm.) deep and ¾-in. (2-cm.) wide. **[Joints with brickwork]**

Form a triangular fillet with polyester filler so that the

coating may run upward into the gap at an angle of 45°. This technique provides a neat and—what is most important—waterproof joint between the coating and the wall, which will even resist pouring resin.

Having finished the coating, you may now seal the remaining triangular gap with polyester filler and smooth the surface by sanding.

Follow the instructions given on pages 79–80 when applying the coating, but lay the dry mat on the primer-treated concrete. Then, impregnate the dry mat by rolling it from the top with a sheepskin roller soaked with activated resin. Roll the mat firmly. It is self-evident that here, too, the overlappings of the single widths of mat are frayed to achieve smooth joints.

Expansion
joints Special care must be taken when coating large areas. Expansion joints in the concrete must not be covered with laminate but must be left to appear in the glass fiber coating. In such a case, you can either coat the whole surface in one run and later cut the expansion joints open with an abrasive wheel or only laminate to the very edge of the joints. Later on, the joints are filled with a permanently elastic joint-filling compound or masked with a permanently elastic jointing tape.

If projecting sections do not bother you, you can compensate for expansion on the principle of an omega-bend, which is used for the compensation of expansion and contraction on pipelines. In this case, you can again use the corrugated cardboard tubing normally used for air ducts, which can be bought as a mold for reinforcing profiles, as already mentioned. Take a length of tubing, cut it lengthwise, and fit the semicircular profile with some polyester filler over the expansion joint. Allow the mat to run over these profiles. This method can also be used for coating large flat roofs made from chipboards.

ROOFS: Polyester and glass fiber are also suitable for repairs to tar-board roofs, which, once they have begun to leak, are everlastingly in need of repair. A standard repair with tar and tar board will only last for a short time. By coating such a roof with glass fiber, you may generally dispense with expansion joints, as the tar boards will stick firmly to the laminate but, at the same time, are not fixed to the roof structure and can compensate for the expansion and contraction of the roof due to heat or cold.

Do not use
primers
containing
solvents The grained and weatherbeaten surface of the tar board allows good adhesion of the laminate and no primer is required.

In fact, the use of a solvent-containing primer would even be fatal, as the solvents would attack the tar board and cause the laminate to change color.

As a rule, a single layer of standard mat will do for sealing and reinforcing a flat roof. If the tar board is still fairly new and only slightly weatherbeaten, a lighter mat, weighing only 1 oz./sq. ft. (300 g./m.^{2}) and requiring less resin, will help to reduce the cost.

Here again, the mat is rolled out dry on the roof and then impregnated with a sheepskin roller. Besides normal standard resin, faster reacting types of resin are also used for this purpose, as they speed up the work. The coated surfaces harden more quickly and can be walked on sooner.

When using one layer of standard mat, you will need approximately 4¼ oz. of resin per sq. ft. (1.3 kg./m.^{2}). Tar board that is very much weatherbeaten, so that it has a rather porous and scarred surface, will require much more resin, so that the consumption may reach a figure of 7.5 oz./sq. ft. (2.3 kg./m.^{2}), which must be allowed for when ordering resin. The completed coating can receive a presealing coat of tinted resin, into which a layer of polyester fiber scrim can be embedded for a better finish. A special type of air-drying polyester resin, LT lacquer, must be used for the final sealing coat; this, too, can be tinted with pigment. To improve the heat reflection, this resin may be mixed with 5 percent of aluminum powder.

Recently, glass fiber has got a competitor as a sealing compound for tar-board covered flat roofs. Instead of polyester and glass fiber, highly elastic two-component PU coatings are now being increasingly used, reinforced by coarse-meshed polyester fabric.

HOW TO BUILD YOUR OWN SWIMMING POOL

Thanks to its specific properties, glass fiber is an ideal material for swimming pools and ornamental garden pools. It is resistant to weathering and most chemicals, as well as to changes of temperature, and has a smooth, nonporous and easy-to-clean surface plus a low thermal conductivity—most desirable for heated pools. It is also easy to work and not too expensive.

Polyester and glass fiber allow every handyman to build a first-class pool—a do-it-yourself job that will last several decades. Most commercial polyester pools are built up from prefabricated laminates, which are molded in female molds and later on joined with bolts and nuts, although some smaller pools are molded in one piece. The do-it-yourself method is based on the principle of coating a dead male mold. The result is hardly inferior to the commercial product and, in some respects, better. A home-built pool is laminated in one piece, which must be regarded as a more appropriate working method for the material involved than bolting single elements together, which is a rather complicated job. It is essential to ensure a perfect sealing at the joints to make the pool watertight and, at the same time, to achieve an even load distribution and prevent point loads—assembly of such a pool is not a do-it-yourself job. Last, but not least, commercial prefabricated pool elements are too expensive for many people. When building your own pool with glass fiber, you can get a fine polyester pool measuring 28 ft. x 15 ft. x 5 ft. (8.5 m. x 4.5 m. x 1.5 m.) for about $1,000 (£500).

Solid working surface

For several reasons, it is not possible to apply resin and mat directly to the soil, as the resin would be absorbed by the soil and the mat would soon lie on the bare ground without bonding. If you have ever tried to do the job in that way, you will have found that the soil is something like a bottomless pit—the resin disappears with the speed of lightning, however many coats you apply. If you try to solidify the soil by stamping—which, incidentally, is an advantage when laminating in the proper way—there will be no change; the resin will disappear all the more quickly due to the capillary attraction of the solidified soil.

Finally, it is essential to insulate the laminate against the moisture contained in the soil, which would cause the polyester to saponify. Insufficiently embedded glass fibers would, by virtue of capillary attraction, allow water to penetrate into the laminate via the fibers and reduce the strength of the glass fiber.

Good ground preparation is, therefore, a prerequisite. It can be achieved in four different ways:

Coated Brickwork

Rectangular swimming pools with vertical sidewalls, which are preferred by enthusiastic swimmers, should have a 5-in. (10-cm.) base of concrete, reinforced with a layer of woven steel fabric. It is cast from lean concrete. The sidewalls of the pool are then erected on this base. They are built up from bricks.

When making the lean concrete base, you must already consider the required slope of the bottom toward the main drain of the pool. Even the drainpipe leading to this drain must be already fitted with the necessary slope.

The finished brickwork is prepared for the glass fiber coating by giving it a thin coat of plaster. At the same time, all angular corners are filleted (use a bottle as a tool for molding the fillets). The upper edges of the sidewalls are also rounded off inside and outside the pool, so that the mat can be laminated around these edges without any trouble.

Brickwork must be plastered

It is most important that any piping, fittings or equipment be installed as you build up the walls. This also applies to the three other systems described later. Also think of installing the cases for underwater lights. All this is necessary to get any fittings and accessories flush with the laminate. Where parts have to be mounted with bolts, you should think of large mountings to be embedded in the brickwork and its plastering to disperse any loads evenly.

All fittings must be flush with the glass fiber surface

Allow enough time for the plastered brickwork to dry out before you apply the primer, which is indispensable for all coating jobs. It will provide proper anchoring of the laminate and, at the same time, seal the pores of the plaster and prevent air bubbles from getting into the laminate. It may happen that two coats of primer are required, as the first coat is quickly absorbed. Here, too, a one-component PU resin, like G 4, is most useful. You must, however, use a trick in order not to exceed the maximum allowance of three hours between application of the primer and laminating. You may either treat the pool's surface step by step, applying primer to 50–100 sq. ft. (5–10m.²) at a time and laminating them straight away, or you can apply primer to the entire surface in one go, providing you can apply the activated resin mixture fast enough to ensure a good bond. The thin, tacky film on the surface of polyester resin that is exposed to the air when hardening (see pages 28–29 for further details) provides a reliable adherence of the laminate.

Rain does not fit in with your plans

In either case, you must be on good terms with the weather gods, for rain at this time is most undesirable. It may cause the

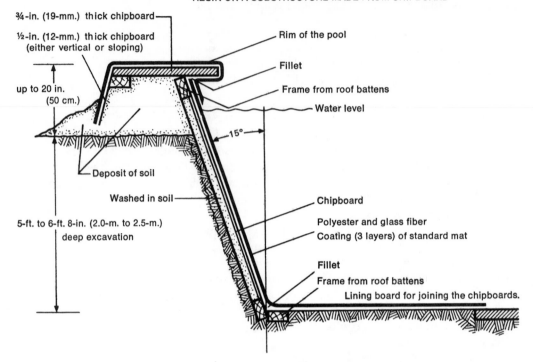

formation of bubbles in the fresh priming coat or saponify the polyester resin, applied as an intermediate coating, which impairs the adherence of the laminate.

This is the reason why it is a good idea to wait for a period of fine weather when building your swimming pool, but since weather has a habit of changing, make sure that you have some watertight film or foil or canvas at hand to cover the pool. Do not, however, laminate in a pool underneath a cover of canvas or plastics, as this may lead to a dangerous shortage of air.

With Chipboard and Inclined Sidewalls

Working with bricks and mortar is not everyone's cup of tea, and many a handyman prefers hammer and saw to trowel and plumbline. Amateurs who are keen woodworkers may prefer a different method of construction. In this case, the substructure is built up from weatherproof chipboards impregnated with phenolic resin, which are nailed down on a skeleton of roof battens.

In the bottom area, lining boards are set flush into the soil, where the chipboards are to be joined. All chipboards should be at least ⅓-in. (8-mm.) thick. The upper edge of the pool is best built in the shape of an inverted U, which gives the whole structure additional strength and rigidity.

Minimum thickness of chipboard ⅓ in. (8 mm.)

In order to make sure that the glass mat will again run smoothly over the upper edges of the pool, you will have to make a fillet in the corner between the sidewalls and the protruding rim of the pool. Some lengths of triangular stripwood are nailed down in the corners of the sidewalls and where the sidewalls meet the bottom of the pool. The edges of the rim are rounded off by treating them with a rasp or an electric sander. Possible gaps between the single panels of chipboard are filled with polyester filler or by glueing thin plywood into the gaps. If they are not too wide, wide masking tape may suffice. Then, apply the obligatory coat of primer. When coating the chipboard structure with the usual three layers of standard mat, complete the sidewalls before starting on the bottom.

IMPORTANT: With this type of construction, it is necessary to have slightly inclined sidewalls to counteract the pressure of the soil. With very solid or loamy soil, an inclination of only 15° will normally be sufficient and is hardly regarded as an impediment by swimmers. With very light and loose soil, the inclination of the sidewalls must be increased (maximum 45°).

As a rule, the excavation for the pool is made approximately 8-in. (20-cm.) bigger all around than required by the overall dimensions of the pool. The remaining gap is filled with soil after the pool is completed. This is best done while the pool is being filled, in order to allow equilibrium between the pressure of the water and the soil and to prevent warping of the walls.

For Ornamental Pools and Free-Form Swimming Pools: Premold with Plaster and Jute

Ornamental pools and swimming pools with irregular outline—and, of course, rectangular swimming pools too—can be laminated on a premolded base coated with plaster and jute. For this purpose, the soil is excavated in the desired shape of the pool and smoothed with a shovel. Then, soak jute, which was cut to the required size, in a thin, pasty mix of water and plaster and lay it down onto the prepared soil. Wrinkles and bubbles are smoothed with one's hands. As long as it is wet, jute fabric can be made to match even spherical shapes, if you stretch and draw it. If need be, some cuts will be a help.

Allow the single width of jute to overlap about 2 in. to 3 in. (5 cm. to 8 cm.) at the edges. The joints can be hidden by giving them a coat of thin plaster mix, which is brushed on with a whitewashing brush. This brush also helps you to seal any parts of the fabric that might not be completely covered with plaster. As it is easier to work from inside the excavation, you had better start by applying the jute to the sidewalls and finish with the base. The thin plaster/jute shell can, of course, not carry the weight of a man without being damaged. Once the bottom is also covered with plaster and jute, you will have to wait an hour until the plaster has turned hard. If the soil is very wet or if the weather is humid, it may even take several hours until the plaster is sufficiently hard and dry. Generally, an inhibitor is added to the plaster mix in order to extend the hardening time and allow some more time for lining the excavation with jute. Fish glue is an excellent inhibitor; this is available from hardware stores and pharmacies and should be added in a concentration of 3 percent of the water required for the plaster mix. Another useful inhibitor is gum arabic, which is also available, as granules, from hardware stores and pharmacies. Above all, it also plasticizes the plaster mix. But even wallpaper paste (cellulose glue) extends the fairly short pot life of plaster. The smoothed and hardened plaster/jute shell is then sealed with a coat of a one-component PU (for instance, G 4), not so much as a primer or bonding agent, since the plaster/jute shell will rot anyway and does not have any structural function. The main purpose is to seal the fine pores in the surface of the plaster and, thus, to make sure that there will be no bubbles in the laminate. As the plaster shell only serves as a laminating base and does not contribute to the strength of the glass fiber pool, the sidewalls of such a pool must, at all events, be sloping, to counteract the pressure of earth. An angle of 45° is a safe standard. Steeper walls are less suitable for this construction.

The laminate is generally built up from three layers of standard mat.

The Fourth Method: Bituminous Paper

Lining the excavation with bituminous paper, which is normally used for road-building purposes, is a cheaper and, from the technical point of view, satisfactory way of preparing the ground for very large pools. This thick brown paper acts as a resin-proof barrier and prevents the resin from trickling into the soil, so that it is most suitable as an insulating layer between soil and laminate. As the paper only contributes very little to the solidity of the soil required for proper rolling of the laminate, it can only be used on fairly compact soil.

Paper prevents resin from trickling into the soil

The excavation is made to fit the desired contours of the pool and is slightly flattened. If possible, firm the soil by rolling it with a garden roller. Then, line the hole with bituminous paper laid down width by width and overlapping at the joints about a hand's breadth. Just as you do when coating a roof, the mat is spread on this surface in the dry state and then impregnated from the top with activated resin. Here, too, you had better work with three layers of standard mat.

This method proved entirely acceptable in the construction of a giant pool in Orlen (West Germany) in 1971. This pool has a length of no less than 110 yd. (100 m.) and an average width of approximately 55 yd. (50 m.), which is as big as a full-size football field. The pool is the largest glass fiber pool in Europe and holds 19,620 cu. yd. (15,000 m.^{3}) of water.

Restoring Old Concrete Pools

Existing pools made from concrete or even plastic sheet can be given a new lease of life with a glass fiber coating, which, at the same time, makes them 100 percent watertight.

Make a test before laminating

Most plastic pools consist of PVC- or polythene sheeting with a thickness of 1/48 in. to 1/24 in. (0.5 mm. to 1.2 mm.). Neither material is attacked by polyester resin. Thin sheeting may, at most, tend to swell slightly, which need not bother us.

Before you start coating such a pool, you should make a test on a piece of sheeting to find out how it reacts to the resin. Above all, you must test whether the underground behind the sheet is solid enough to allow you to roll the laminate properly.

Do not use a primer, as this must inevitably destroy the sheeting, which, in any case, acts as a resin-proof barrier, like the bituminous paper.

Plastic pools usually have sloping sides, which are also necessary for a glass fiber version. If a rectangular pool built up from bricks or concrete is fitted with a prefabricated plastic sheet lining, you had better remove the sheet to apply the glass fiber directly on the plastered brickwork or on the concrete, as described on pages 89–90.

The same method of working is recommended for repairing cracked concrete pools. Here, all cracks are first scraped out with a chisel before they are treated with a primer and filled with polyester filler, which is sanded flush with the concrete surfaces later on.

Repairing leaky pools

If the pool is only leaky or frost-cracked, a single sealing layer of standard mat or 2 oz./sq. ft. (600 g./m.^{2}) mat will do, although it is a bit more difficult to handle due to its greater thickness. But you may as well use two layers of 1.5 oz./sq. ft. (300 g./m.^{2}) instead. Bare concrete pools are given a coat of primer and can then be fiberglassed in the normal way. Pools that were painted with chloroprene paint can be coated straight away, if the coat of paint is still intact and still adheres firmly to the concrete. Wherever the paint is loose or has flaked, the bare concrete must be treated with a primer (make an adhesion test).

Large pools are coated with two layers

Large pools should be given two coats of mat if possible. If the glass fiber coating must also take up stress because the pool is not at all or not sufficiently reinforced, you should use at least three layers of standard mat. As it might be difficult for many laymen to judge the technical requirements, the advice of an expert is extremely valuable.

Sealing and Filling the Pool

All glass fiber pools built by one of the methods described above must receive a presealing coat with tinted resin, which can be improved by embedding a layer of polyester fiber scrim for improving the finish of the surface. Then, the final sealing with tinted polyester air-drying resin is applied. The amount of pigment added to the air-drying resin must not exceed 5 percent, the resin for the presealing coat is mixed with 15 percent to 20 percent pigment.

Observing the minimum temperature of 64.4° F (18° C), allowed for the hardening of the air-drying resin within thirty minutes, is another most important point. Otherwise, the resin will not become sufficiently resistant to the water in the pool, so that the resin will saponify on its surface. This leads to a whitish film, which greatly detracts from the appearance of the pool. Remedy: Empty the pool, sand the whitish film off and apply a new coat of air-drying resin on a warm day. Before the pool is refilled, you should allow at least three warm, sunny days to ensure a sufficient cure of laminate and sealing coats.

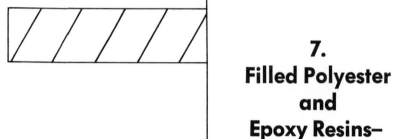

7.
Filled Polyester and Epoxy Resins— Fillers for Any Purpose

Versatile polyester products include the fillers, which were already mentioned in connection with mold making and as primers for metal surfaces to be coated with polyester and glass fiber. The main field of application, however, is, no doubt, the repairing of car bodies, for polyester filler hardens very quickly all through and can generally be sanded within fifteen to twenty minutes, so that a damaged car body may be ready for painting within a few hours.

As good-quality polyester fillers stick extremely well to metal and, at the same time, are highly resilient, you can even fill deep or jagged dents. A good filler even allows you to fill dents up to a depth of 2½ in. (6 cm.). The filler will not come loose later on if you carefully sand the whole area to which the filler is going to be applied until the bare metal is visible.

Resin, Filler, Hardener

Polyester filler is generally supplied in cans, with the hardener coming in a separate plastic or metal tube. But there also are fillers that consist of three components: the liquid resin, filler powder and hardener, which allows you to mix your own filler in the required consistency by adding more or less filler powder. Such systems, however, require some skill to achieve a homogeneous mix. Moreover, an excessively thin mix, which might appear advantageous for special purposes, such as the filling of narrow joints, will also affect the shrinkage of the filler compound, as the fillers added to the resin not only add body but also reduce shrinkage and contribute to firm adhesion of the filler compound by preventing it from shrinking away from the base. Therefore do-it-yourself people are generally better off if they use ready-mixed filler pastes to which only the hardener has to be added. Quite often, the hardener is tinted in a bright red color, which allows you to ensure homogeneous mixing with the generally light gray filler. Only white filler paste is an exception. It is a special product, offered by only one manufacturer, for the repairing of white polyester boats or light-colored polyester furniture that does not require painting after the repair, or is at least easier to paint if white filler is used. White filler must be mixed with white hardener paste in order not to spoil the white color. Three percent is the normal quantity of hardener to be added both to white and gray fillers. The filler and hardener are mixed on a clean, solid plate (for instance, a piece of plywood covered with plastic film or foil). Make sure that, as far as possible, no air is introduced into the filler during mixing. Always use a clean spatula that is free of hardener or already activated fillers for taking more filler out of the can. Otherwise, you will contaminate the fresh filler paste in the can with hardener and, thus, start the polymerization process: the filler paste gradually starts hardening and will be useless after some time.

Ready-for-use mixtures are more practical

Different Qualities

There are fairly great variations in the quality of the various products in respect to a number of points: more or less easy sanding; more or less good adhesion; proper hardening of thin layers; softness and easy application on the surface; fineness of grain, which is most important for the finish after sanding; the tackiness of the surface of the filler after hardening and the specific gravity of the filler, which has a great influence on costs. Of late, top-quality fillers are being produced under vacuum, which results in especially creamy and easy-to-apply fillers.

Spray guns with large nozzle There also are polyester spray fillers. These liquid fillers are mixed with hardener and applied with a standard spray gun. Unlike standard polyester filler paste, spray fillers contain solvents and cannot be applied in coats of unlimited thickness, as you must allow the solvents to volatilize.

The fairly high viscosity of spray fillers calls for wide spray nozzles—a 5/64-in. to 3/32-in. (2.0-mm. to 2.5-mm.) diameter. It is self-evident that to keep the spray gun in working order, it must be cleaned before the filler has cured.

Polyester spray filler can be used for treating both bare and primed metal and wood or glass fiber surfaces. Good-quality spray fillers will cure even in thin coats and have a nontacky surface.

Fields of Application

Let us now return to polyester filler pastes. Besides the various applications already dealt with, these fillers provide a great variety of applications for the practical amateur and the spare-time artist. They can be used for making casting molds for low-melting metals, such as lead and tin. For this purpose, the molds should be tempered in a domestic oven before use. You can also use polyester fillers for weather-resistant sculptures, the supporting skeleton of which is made from thick wire (for instance, thin reinforcing steel). The fairly short pot life of polyester fillers, however, only allows a small quantity to be mixed at a time. If you want to make bigger sculptures, you are better off with special polyester sculpturing compounds, which are very similar to polyester fillers but cure a little more slowly to allow the sculptor more time.

Mechanical retouching is possible Once hardened, these polyester compounds, as well as polyester fillers, can be shaped and retouched by scraping, filing, milling or sanding, which opens a wide scope of styling methods. Even sgraffito techniques are possible.

Tinting fillers is rather a problem job, even though there

are suitable color pastes. But the mixing of the color paste with the filler is quite difficult without the help of special tools. If you want to use substantial quantities in a color other than gray, you should try to contact the manufacturer, who can possibly make a special mix for you. This will inevitably mean additional costs —if you are lucky enough to find a manufacturer at all who is willing to supply you with special mixes in fairly small quantities. Anyway, it will be cheaper to spray or paint the finished sculpture Painting with enamel that dries with a mat finish. This enamel finish will adhere firmly to the hardened filler, which is demonstrated by the use of filler for repairing car bodies. As a matter of fact, not all makes and types of polyester filler allow stove enameling: a fact that is more important where car body repairs are concerned than with sculpture. In case of doubt, you had better ask the manufacturer or supplier of the polyester filler, if you want to have your repaired car body or stove enameled. Good polyester fillers withstand stove enameling temperatures of approximately 194° F (90° C), which is very often marked on the can. If there is no such notice, make inquiries.

GLASS-FIBER FILLERS

There is also another type of polyester filler, the usual mineral content of which is replaced by chopped glass fiber strands. Such fillers are widely used for coating and reinforcing sheet metal, wood and brickwork and thus differ very much from the normal fillers mentioned above.

For the hobbyist, glass fiber fillers are especially useful for repairing rusty car bodies.

For mending rusty car bodies

Sheet metal parts rendered thin by corrosion and parts of the car body already dotted with masses of tiny holes can be restored with a 1/32-in. to 3/64-in. (0.6-mm. to 1.2-mm.) coat of glass fiber filler after careful sanding of the surfaces. The filler contains a thixotropic agent, so that it does not run on vertical surfaces nor trickle down from horizontal surfaces when you have to work overhead. Mixed with hardener, the transparent filler has a pinky yellowish color. It can be applied with a spatula but may also be brushed on, especially if a surface cannot be reached with a spatula and is even less accessible for laminating. If necessary, you can apply a 3/16-in. (5-mm.) coat of glass fiber filler for extra strength.

If the holes in the car body are, however, bigger than a

pinhead, it is better to laminate a layer of glass mat against the rear side of the part, as this provides more strength due to the higher content of glass fibers. In such a case, the part must be sanded to the bare metal and is then given a thin coat of polyester filler, before the surface is brushed with activated resin. Then, the dry mat is laid in place and fully impregnated from the top side by dabbing it in place with a brush soaked with resin. If the holes are bigger than peas, we recommend the method described on pages 155–158 and that you preimpregnate the mat on a piece of plastic film or foil in order to apply it already soaked with resin.

The surrounding less-corroded surfaces can then be reinforced with glass fiber filler.

For modeling and repairs in the house Glass fiber filler is also very interesting for model makers, who can give the engine compartment of their model aircraft or boat a fuel-proof coating, which, at the same time, adds strength. Moreover, glass fiber fillers can be used for quick repairs on model fuselages and model boat hulls. It also fills cracks in brickwork, protects the wooden frames of verandas against moisture and adds years of wear to materials that themselves have little resistance to abrasion. You may also use it for sealing leaky tubes, unless these are subjected to pressure, or for mounting an anchor in a wall.

POLYESTER AND EPOXY RESINS WITH METAL OR CERAMIC POWDER FILLERS

The thought of repairing metal parts is terrifying to most amateurs, who generally lack the proper tools required for a workmanlike repair. Welding or hard soldering are techniques generally reserved for experts, because these methods of joining metal parts require skill and experience if the result is to be satisfactory.

Anyhow, there is no reason for despair, if a drilled hole collapses, a tube becomes leaky, a seam breaks, a thread is damaged or any metal part begins to show signs of wear. Any such damage can be fairly well repaired with easy-to-use repair compounds based on polyester or epoxy resin containing fine metal powder. Such materials can be bought as complete kits and are even available in small quantities to meet the requirements of an amateur.

Properties and applications depend on the type of resin and the added filler, which is supplied in a graduated grain size. Suitable filling materials are aluminum, iron, copper, steel, brass and ceramic powder.

Resin and filler form special compounds, the properties of which are governed by the two components and which are more influenced by either the resin or by the filling agent, if the system allows a variation of the mixing proportions. Besides resin/powder systems, you may find two-paste systems, consisting of a ready-for-use mixture of resin and filler, plus hardening paste in separate tubes. Such systems are the exception; most repair compounds are of the liquid pure resin plus filler powder type, with the filler also containing the hardener.

Choosing the Resin Base

Most companies offer both polyester and epoxy-based systems. Polyester systems are much cheaper and less critical as far as the mixing proportions are concerned, for they provide many starting points for polymerization. They harden quickly (in less than an hour) but shrink up to 3 percent by volume, even if you choose a first-class product. The danger of heat accumulation must also be considered, as it will increase the shrinkage. Therefore, thick coats should not be cast all at once but built up layer by layer, allowing each coat to cure before the next one is applied. The two resins also differ in their mechanical strength.

Polyester or epoxy resin?

Polyester-based compounds are more resistant to crushing but are fairly brittle. Epoxy resins are tougher. Both systems are by no means completely cured when they appear to be hard. The hardening generally goes on for several days and even weeks. The process can be accelerated by curing the piece in a domestic oven for about one hour at a temperature of 212° F to 248° F (100° to 120° C) once the resin appears to be hard. Avoid higher temperatures, as they are detrimental to the resin.

With regard to their heat resistance, epoxy compounds excel polyester compounds. Polyester compounds are heat resistant within a scope of −58° F to 356° F (−50° C to 180° C), while epoxy-based compounds resist temperatures of about 356° F (maximum temperatures 392° F to 482° F [200° C to 250° C]). As a matter of fact, you must put up with a reduced mechanical strength of the material under such high thermal stress. Epoxy resin is also superior with regard to adhesion, so that special care must be taken when using polyester compounds for special treatment by roughening, sanding, notching, drilling inclined anchoring holes or by inserting hedgehog-type metal studs into the surface of the work piece in order to achieve a durable joint. Enlarging the contact surfaces is also a help where possible.

Another point in favor of epoxy resin compounds is their better chemical resistance to many substances. This comparison demonstrates the superiority of epoxy-based compounds. In most cases, however, the amateur will succeed with the less-expensive polyester compounds, especially because their quicker hardening makes less demand on patience. Only if extreme stress is envisaged or in case of extremely difficult working conditions should you prefer epoxy resins to polyester compounds, especially because a clever application can compensate for many a disadvantage of polyester compounds.

Epoxy compounds —expensive but more resistant, polyester compounds cheaper and quicker

It also benefits the results if you arrange the repair in a way that makes the best use of the advantages of polyester resin. You can, for instance, take care that a repair is only to compressive stress, as polyester has outstanding qualities in this respect. If this is not possible right from the start, you may use external steel casings or collars to overcome the poor impact strength of polyester resin. Quite often, it will be enough—as, for instance, by embedding woven glass fiber, which then receives a thick coating with a mixture of resin and powder.

The following table may help you in deciding between epoxy or polyester compounds. It shows the advantages and disadvantages of both types of resin used in repair compounds (1 = superior, 2 = inferior).

CRITERIA	POLYESTER RESIN	EPOXY RESIN
Low price	1	2
Curing time	1	2
Wider margin in mixing proportion	1	2
Adhesion	2	1
Compressive strength	1	2
Impact strength	2	1
Shrinkage	2	1
Chemical resistance	2	1
Heat resistance	2	1

Properties of the Compounds

Metal-filled liquid plastic resins are marketed under different names, such as cold metal, plastic metal, plastic welding compound and metal filler. The amateur should regard these names and specifications only as hints to how these materials can be used. With regard to the properties of these materials, the emphasis lies much more on "plastic" than on "metal," as the properties of the resins that are mixed with different kinds of metal powder are to a greater extent controlled by the synthetic resin. This is quite evident if you realize that metal-filled resins do not show as distinct differences as the active metals. Thus, cured metal-filled resins are all poor conductors of electricity. Their content of metal only improves the electrical properties as far as the specific electrical resistance of the resin—which is comparable to that of an insulating material as far as pure resin is concerned—will reach the level of high ohmic resistors, if a large quantity of metal powder is added to the resin. The content of metal powder, consequently, does not transform the nonconductive plastic into a material with the electrical conductivity of metals.

The same applies to thermal conductivity. Plastic resins, which are known as poor conductors of heat, only become moderate conductors of heat by adding metal powder, but their thermal conductivity can by no means be compared to that of pure metals.

Comparison of the adhesive properties, on one hand, and the rigidity, hardness, compressive strength and durability of filled and nonfilled resins, on the other hand, indicates a genuine compound material. The addition of fillers reduces the adhesion of epoxy resins by about 30 percent and of polyester resins by up to 50 percent, compared to pure resins used for two-component glues, which have an astonishing tensile strength ranging from 2,750 lb./sq. in. to 4,812 lb./sq. in. (200 kg./cm. to 350 kg./cm.) for epoxy glues. Unlike the adhesion qualities, the stiffness, hardness, compressive strength, durability and shrinkage resistance of polyester resins are improved by adding metal powder.

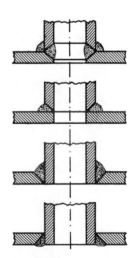

Increasing the contact area improves the strength, as it does in welding

Choosing the Filler Material

Various points of view have to be considered when choosing the filler. When repairing metal objects, the finish is often the main point. Therefore, you will probably use the filler that is most similar to, or even identical with, the kind of metal of which the object to be mended is made. In such cases, the

Heat
resistance

Sliding
properties

Repairing
cracks

"cosmetic" effect holds the spotlight. If a repair is exposed to heat, you have to find means of improving its heat resistance. In such a case, iron powder is the most suitable filler.

Damaged or worn-out bearings are best repaired with resin and aluminum powder, which excels by fairly good sliding qualities, and which may even be further improved by adding up to 10 percent of molybdenum disulphide or graphite powder. Such additives can, however, only be added to polyester resin-based compounds.

For good insulating properties, nonmetallic fillers are recommended, such as ceramic powder, which can also be tinted in any desired shade of color by adding dry earth color powder obtainable from a paint store. You can add up to 10 percent of color powder.

Even with other kinds of repairs where the insulating properties do not matter, ceramic fillers prove to be quite useful. In a job done by the author, the fly cutter of an electric coffee mill, which had come loose after the original thermoplastic hub had broken, was satisfactorily mounted on the shaft of the motor with a polyester-based ceramic repair compound (omni-PLUS type K-KM). Owing to the extreme hardness and very good resistance against abrasion, the repair does not show any signs of wear after more than two years of daily use. The cutter is still firmly fixed to the shaft.

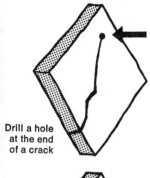

Drill a hole
at the end
of a crack

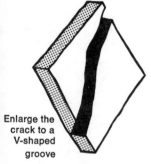

Enlarge the
crack to a
V-shaped
groove

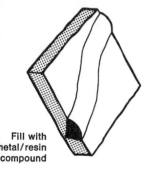

Fill with
metal/resin
compound

Some Points Worth Considering

Correct preparation is most important for repairing metal parts. Above all, the parts must be free from grease and dirt to allow the resin to anchor in the material. In addition, it is advisable to enlarge the contact zone as far as possible, which is even more important with polyester compounds than with epoxy resins. The larger and rougher the contact surfaces, the more durable the repair will be. V-shaped widening of cracks in massive metal parts (motor casings) is a practical way of increasing the contact area. You may grind the V-shaped groove to a depth of two-thirds of the total wall thickness of the part to be repaired. It is appropriate to drill a hole at the ends of the cracks before the filling compound is applied. Thus, you achieve a round end, which is less susceptible to an extending of the crack. Then, the crack is widened by filing or sanding to a V-shaped groove.

Enlarged or damaged drill holes or threads must be drilled

open to at least one and one-half to twice their original diameter. To allow the repair compound to anchor firmly in the hole, you should strive for a mechanical grip by filed-in grooves, coarse threads cut into the hole or little anchoring holes drilled into the wall of the hole. Once the repair compound has hardened, the required hole can be drilled with a sharp twist drill. For this purpose, you should use a hard metal drill. Finally, you may even cut a new thread into the drilled hole. If the thread must withstand high-tensile loads, you had better use a threaded metal bush, which should have a collar on the side opposite to the direction of the tensile loads, so that this threaded insert may transmit the tensile loads directly on the casing. Drill or mill the casing to match the collar of the insert. A fairly wide gap between the bush and the enlarged drilled hole in the casing is then filled with a metal/resin compound. Repairs to worn surfaces will need increased adhesion by mechanical aids, such as inclined holes drilled into the basic material, cutouts tapered from the back, dovetailed grooves or protruding bolt heads. The wide tolerance in proportioning polyester-based compounds allows you to adjust these materials to nearly any consistency. The scope ranges from a pourable mix with little filler, which only gives the resin the desired color, up to thick pasty mixes, which are thick enough for kneading them into shape. Thus, you may adjust your mix according to the special requirements of the repair. But you must keep in mind that very little filler will promote the shrinkage.

Any desired consistency with polyester compounds

This applies also to the use of thinners, up to 10 percent of which may be added to some polyester-based products, such as those offered by Omni-technik. As the thinners do not remain in the compound, but volatilize, the quantity of the material is reduced, which leads to further shrinking.

When repairing machine parts, it may happen that for special reasons (surface structure, adhesion problems, etc.) fairly thin or only slightly pasty mixes are required, which call for special means to prevent the mix from running. In such a case, the area to be repaired must be surrounded with a little dam of Plasticine or wax. You should, however, take special care that these masking materials do not come into contact with the surfaces to be coated, as this will inevitably cause adhesion problems. Standard self-adhesive tape (Sellotape) or special creped masking tape can also be used. In order to achieve a perfectly smooth surface, it is advisable to apply a slightly thicker coat of repair compound than is actually required and to sand

Masking with Plasticine or wax

the hardened material with carborundum paper until it is perfectly flush with the original surfaces. This even enables you to restore proper sealing joints (contact surfaces).

Castings from Resin and Metal Powder

Metal-filled plastic resins are of special interest for do-it-yourself or modeling work as they can be used for molding fans and pump impellers, casings, model guns for scale-models of historic ships and similar parts and thus avoid the problems of metal casting.

Semiliquid plaster mixes are suitable for molds, especially for small masters. Once hardened, the surface of the plaster mold must be sealed with DD lacquer. Other mold materials require commercial release wax, silicone-containing sprays or, if need be, even oil containing molybdenum disulphide. Even simple floor polish will do. You may completely dispense with release agents, if you use a mold made from silicone rubber, which is absolutely necessary if you want to cast masters with undercuts.

If the casts are to be painted later on, it is advisable to sand their surfaces after hardening and to avoid the use, if possible, of release agents containing silicone. Before any paint is applied, the thin film of release which is left on the casts must be washed off with solvents. Then, sand the cast slightly, remove any dust and apply a coat of commercial enamel. In most cases, however, it is better to dispense with painting altogether.

OTHER FILLERS FOR EPOXY AND POLYESTER RESIN

There are other fillers that can be added to resin and open a wide scope for experiments.

• WOOD. Wood meal is a popular filling material for polyester casting resins and is added in a quantity of 15 percent to 30 percent. Such a mix, based on weight proportions, is still liquid enough to fill even very fine grain structures of the mold. If you want to imitate wood, the cast should be taken out of its mold as soon as it is hard enough and then brushed with a solvent (for instance, acetone or methylene chloride). Thus, the surface of the cast is attacked by the solvent and becomes slightly porous, so that you may later treat the cast with stain to give it the natural color of wood.

Apply the stain with a brush and work line by line with different shades of stain to produce a woodlike finish. This enables you to achieve more realistic structures than those obtainable with tinted resin or wood flour.

The more wood flour added to the resin, the more easily will the casts take up the stain, but you will have to pay for this advantage with poor flowing characteristics of the mixture and a greater risk of bubbles. A further disadvantage of only adding wood flour to the resin is the fairly high weight of the cast, which inevitably reveals that the part is not made from wood, even though the cast shows the exact grain and pores of the wooden masterpiece from which a silicone rubber mold was taken.

• MICROSPHERES. There is a means of avoiding this particular weight problem: microspheres made of glass, phenolic resin or ceramics. Their specific gravity is about 13.73 lb./cu. ft. (0.22 g./cm.^{3}), and the grain sizes varies between 10μ to 250μ, which is 411,000 in. to 1/100 in. (1/100 mm. to ¼ mm.). (These figures apply to sodium borisilicate glass spheres as used by the furniture industry.) Each of these tiny glass spheres has a wall thickness of 2μ (8/1,000 in. [1/500 mm.]). The special grain size distribution leads to an optimum packing, as many small spheres will settle between the bigger ones. The fine hollow spheres look like dry sand. They are colorless and perfectly round.

They are easily wetted with resin and allow highly filled mixtures that are still flowable. Microspheres are nontoxic and have no chemical effect on the curing of the resin. Even under the heat of the exothermic hardening process of the resin, they do not release any gas, so that the risk of bubbles or a faulty surface is very small. The very low specific gravity keeps down the weight of polyester resin mixed with microspheres and allows you to reduce the specific gravity of polyester resin—68.67 lb./cu. ft. (about 1.1 g./cm.^{3})— to that of wood—53.07 lb./cu. ft. to 59.31 lb./cu. ft. (0.85 g./cm.^{3} to 0.95 g./cm.^{3}).

At the same time, the resin that was mixed with the microspheres gets the mechanical properties of wood, so that it can be sawed, drilled, screwed and nailed, which is most astonishing if you remember how brittle polyester resin is by its nature.

As you may imagine, there is a "but" in the case. At first **Expensive** sight, microspheres are not cheap at all. But the fairly high price **but economical** per pound is deceptive, for one pound of microspheres is equivalent to about 4.3 pints (1 kg. = approximately 4.5 liters). As the spheres add volume to the resin, less resin is required, which again reduces the costs.

Moreover, there are some more advantages if you compare

microspheres to the fairly heavy mineral fillers. Compared with a mixture containing the same volume of fillers, a mass filled with microspheres is much lighter. If you compare two mixes having the same strength, the one with microspheres is again lighter and stiffer. Even the weight to strength ratio is better in masses containing microspheres than those containing the same volume of fillers. Above all, the addition of microspheres reduces the shrinkage of polyester resins, makes the cast easier to remove from the mold, and improves the impact resistance of polyester resin more than conventional fillers. Finally, casts made of a mixture of resin and microspheres are slightly elastic and resilient, which even allows tightly fitting plug-in joints. Above all, they also reduce the tendency of the molded parts to warp or shrink during aging.

Perfect imitations of wood
This catalog of good properties opens a wide field of applications for microspheres. Furniture works often use them for imitation wooden panels and achieve excellent results. They quite often manage to mold imitations of wood carvings which are so true to nature that it is even difficult for experts to say whether the carved patterns on the doors of a cupboard are real wood or plastic castings. A well-done finishing with stain and clear varnish adds very much to realism and the illusion of real wood. This is a wide and most interesting field for your own experiments!

The following recipe may serve as a hint for your own experiments. It is based on the use of IG 101-type microspheres, as supplied by Emerson & Cuming Inc. Mix, by weight:

100 parts of polyester resin and hardener
 40 parts of wood flour and
 15 parts of microspheres IG 101

If necessary, add some aerosil paste for thickening and for a more even distribution of the microspheres.

• STONE. Just as you can simulate wood by adding wood flour to polyester resin, you can simulate stone or marble with polyester resin serving as a binder. Work in a well-waxed mold made from tin or melamine-coated chipboard that is fitted with a frame of plywood lined with Formica strips.

Having poured a thin gelcoat of clear resin into the mold and having allowed it to cure, you take marble dust, colored stones in a graduated grain-size distribution or even dry arenaceous sand to which 15 percent to 20 percent of polyester resin mixed with hardener are added. Mix thoroughly and pour this mixture into the prepared mold, distribute evenly with a spatula and solidify by rolling it with a length of tubing or a rolling pin wrapped in a plastic sheet. You may even stick bigger decorative stones directly onto the hardened gelcoat. Use clear embedding resin (see page 115) for glueing these parts in place, and then fill the mold with a mix of resin and fine-grain stone material similar to that described above. Broken marble pieces are ideal for this purpose.

Homemade artificial stone slabs can be used as window sills, as decorative panels for flower boxes or troughs or lining the walls in a hall or in the bathroom. Last, but not least, you can use them as tops for small tables or pedestals for flower pots.

Bigger tabletops should be reinforced by embedding a layer of light woven steel fabric, which should be placed in the middle of the thickness of the cast. You can fix threaded brass bushes to the crossings of the woven steel fabric by welding or hard soldering, which may later serve for mounting the tabletop on its support. Do not use brass straddling dowels for this purpose, as they will grow thicker when the bolts are tightened and cause the cast to crack. Bigger plates should be stiffened lengthwise by bolting two lengths of stripwood, flat or L-shaped iron bars to it from underneath, which may also serve as mountings for the legs of the table.

The surface finish of such a cast tabletop turns out to be perfect and glossy if the mold had such a good finish, too. But you may even improve this finish by polishing the surface with polyester polishing paste. As polyester resin is fairly brittle, tabletops made from this resin can be damaged by impacts, so that you might possibly be better off by using epoxy resin for a homemade coffee table that is going to be used very often. But you can also use one-component PU resins as a binding agent (see page 135 for further details).

Epoxy resin for a coffee table

To save materials and costs, you will, of course, not cast the whole thickness of the tabletop from resin and decorative stone material. If the underside of the panel is not visible, a supporting coat can be made from dry quartz sand with a granular size between 0 in. to 9/32 in. (0 mm. and 7 mm.) and fine quartz sand mixed in equal parts and bound with 10 percent polyester resin added to the filling material.

If the tabletop is exposed to strong ultraviolet light (sunlight), it is advisable to use light-resistant resin for the gelcoat and the decorative layer immediately underneath, in order to prevent subsequent color changes due to yellowing of the resin.

Even coarser decorative fillers can be used for casting resin-bonded tabletops. However, you will generally not be able to avoid concentrations of resin between the rather bulky bodies of the filling material, so that you must use special types of resin that produce little stress when hardening. Apart from the fairly expensive epoxy resins, some polyester resins (for instance, a clear-casting or embedding resin) will meet these requirements.

You can use standard polyester resin combined with the small pebbles sometimes used in flowerpots. You can buy such pebbles cheaply in a flower or gardening shop. For casting a plate measuring 14 in. × 14 in. × ⅝ in. (35 cm. × 35 cm. × 1.5 cm.) you will need about 5.5 lb. to 7 lb. (2.5 kg. to 3.0 kg.) of pebbles and about 1 lb. (½ kg.) of polyester resin. A waxed baking pan (no longer in use) can be used as a mold. The pebbles are dried in the oven and then evenly dispersed onto the gelcoat, previously brushed onto the baking pan and now hardened. Consolidate the pebbles by slightly shaking the pan. Then, mix polyester resin with the required quantity of hardener and cast it over the pebbles; the entire surface should be brushed with resin to achieve a glossy finish and better adhesion. Standard polyester resin is fluid enough to fill any cavities between the pebbles without the risk of the formation of bubbles. The cast panel can be removed from its mold after proper hardening (approximately three hours). It has two different surfaces: one is entirely smooth with a high gloss where the gelcoat was applied to the mold; the opposite side shows an irregular surface, with the pebbles not completely embedded in the resin. But this side also has a high gloss. Make your own choice of which side you prefer for the top or which one is better suited for the special purpose for which the plate was cast. If it is intended as a support for a vase or a statue, you should take the flat side for the top side, to provide a safe support. It is quite attractive to embed broken colored glass, cuttings from nonferrous metal, mussel shells or colorful stones in resin. In such a case, you should, at all events, use clear resin that is stable to light and produces limited stress when hardening, as larger islands of pure resin cannot be avoided between the single objects if you want to achieve an effective and decorative pattern. The resin may be used in its original crystal clear form or tinted in light transparent colors. But you must be most careful when adding pigment to the resin to make sure that it remains transparent and is not rendered opaque. Besides tabletops, friezes and decorative panels lighted from the rear lend themselves to this technique. You can even make impressive lampshades from several panels cast in this way.

Thickening Agents

While on the subject of additives for polyester and epoxy resins, reference should also be made to thickening agents used to modify the resin in order to prevent it from running down vertical surfaces and to increase its viscosity on a hot day in order to ensure proper laminating. The very light aerosil powder or aerosil paste is generally used for rendering resins thixotropic. Aerosil paste is, in fact, aerosil powder mixed with resin. This paste is favored by many amateurs, as it is easier to add to resin. Due to its very low specific gravity, aerosil powder does not mix very well with resin. If you are not clever, you may manage to disperse more white powder in the air than in the resin. If you are working in the open air and if there is even a little wind, aerosil powder is a dead loss—in the truest sense of the word.

The best way to mix aerosil powder with resin is to prepare a thick paste with only a little resin and then disperse this paste in the full quantity of resin, stirring vigorously. The normal quantity of aerosil powder to be added to the resin is between 1 percent and 3 percent. Aerosil paste is supplied in different grades of concentration. As a rule between 10 percent and 30 percent is to be added to the resin, depending on how much the resin should be thickened.

Resin-bound Mortar

It is also possible to mix a mortar from dry sand and polyester or epoxy resin. Polyester resin is, however, rarely used due to its brittleness. Epoxy-bound mortar is better than polyester mortar due to its higher elasticity and superior adhesion. But it is fairly expensive, even though a percentage of 10 parts to 20 parts of resin and 80 parts to 90 parts of dry sand (based on the weight) is sufficient.

Recently, epoxy mortar has lost some of its dominant role, at least as far as the do-it-yourself field is concerned, where it admittedly has never played a very important role, as one-component PU resins have proved to be at least equally suitable for making resin-bound mortar. One-component PU resins are very fluid, easy to use and fairly cheap, compared with epoxy resin, and also have excellent bonding properties and durability.

EPOXY FLOORS FROM A CAN

Filled epoxy resins are offered for coating the floors of kitchens, bathrooms and terraces. The activated resins are poured on the well-scrubbed floor, where they are evenly dispersed with a comb-spatula. They can be applied by skilled amateurs, but it is essential to adhere strictly to the instructions of the manufacturer.

Resistant but slippery Epoxy-coated floors are very resistant to mechanical and chemical influences, but they may also turn out to be quite slippery, especially when wet (bathrooms). It may be better, therefore, to choose special types, which produce a slightly grained finish, or resins that can be made nonslip by scattering some fine dry sand on the not yet hardened surface.

The proper choice of type of resin or flooring compound can be quite difficult, and the amateur would be well advised to consult an expert to avoid failure with the not very cheap material.

Porous, oily floors and, also, floors that are sometimes wet mean special risks. But there also are special types of epoxy resins that are not affected by a wet base. Oily floors must, at all costs, be cleaned with caustic cleaners, such as Spic and Span, P 3 or oil-dissolving chemicals, in order to ensure a proper adhesion of the resin coat. Porous undergrounds must be given a sealing coat of primer before application of the resin.

Sloping floors may cause special problems as some flooring compounds are self-leveling, so that they will run off in the direction of the slope. If you cannot level the floor before coating it, because a certain gradient is required to guide water to a drain, you can only use coatings that have to be applied with a spatula. As a matter of fact, they are very difficult to apply, so that an amateur may run into trouble.

There are solvent-containing epoxy coatings and others that do not contain any solvents at all. The latter can be applied in thick coats in one run without running the risk of the formation of bubbles or poor hardening.

Flooring compounds that contain solvents must, as a rule, be applied in several not-too-thick coats in order to allow the solvents to volatilize. With systems that do not contain solvents, you can apply the whole quantity of the compound onto the floor, and the resulting coat will not become thinner due to the loss of solvents, which has to be envisaged with solvent-containing systems; this is a point that should be considered when comparing prices.

By the way, two-component flooring systems need not be free from solvents. Therefore, you should carefully study the leaflets and labels on the cans. There also are systems that can be diluted with water—so-called epoxy emulsions; they are recommended for wet floors.

Carefully study leaflets or labels

Besides the epoxy system dealt with in this chapter, you can also buy flooring compounds based on other types of synthetic resins, such as PVC, PU or acrylic resin; some compounds even consist of a mixture of different binding resins. Some of these compounds are one-component systems; others consist of two components.

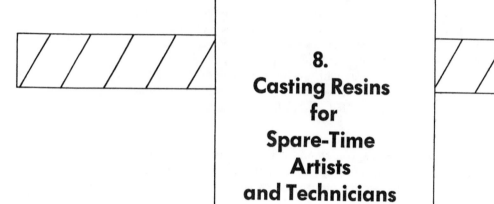

8.
Casting Resins for Spare-Time Artists and Technicians

Coins, flowers, seashells, sea horses, starfish or other small ornamental items, which can be embedded in clear resin, are often admired in souvenir shops and boutiques and make quite popular gifts. Many people are fascinated to see the embedded object in brilliant colors—maybe even enlarged by the special shape of the glasslike plastic block—and be well aware at the same time that it cannot be spoiled by touching.

The makers of such objects do not hesitate to ask quite a bit of money for these visually and psychologically attractive objects. Quite often, you will be asked to pay $5 to $10 (£3 to £4). For about the same amount of money, you can also buy a complete kit containing casting resin, hardener, molds, objects for embedding, mixing pots, pigments, and so forth or buy all these items separately. Instead of only one object, you can get enough material to make several castings. The fun you enjoy by your creative hobby is supplied as a free extra.

Not long ago embedding was not only a fairly expensive proposition but even a most intricate job. For many years, only expensive epoxy resins could be used for embedding, as only such resins proved to be sufficiently fast to light and crystal clear. At the same time, they only caused little warping when hardening.

Now there are also polyester resins that meet these requirements and enable nearly everybody to take up this hobby, especially as these new polyester resins are by no means difficult to use and—what is maybe even more important for the sometimes impatient leisure artist—harden much more quickly than the epoxy resins previously used.

It is most important that you only choose special embedding resin if you buy the resin separately. If, however, you buy a kit, you may forget worries about the question of which resin is suitable.

Embedding resins must be fairly liquid in order to be able to enclose even finely detailed objects, and, also, as already mentioned, be crystal clear and light resistant. The former need not apply at all events to the still liquid resin but is essential to the optical qualities of the hardened block.

Change of Color

There are, however, some types of resin that are slightly greenish in their liquid state. Once the hardening reaction has started, they become as clear as water, without a trace of color. This loss of color is quite useful for the purpose of casting, as it enables you to tell how long you may still try to remove air bubbles from the still liquid casting without worrying about permanent marks. It also indicates how long the reactive mixture can be moved. Above all, the changing of the color announces the instant when the surface that is exposed to the air should be sealed with a piece of glass or Melinex or Mylar sheet in order to achieve a nontacky surface. If, however, you cover the resin too early, you will run the risk of entrapping air bubbles, which may still rise from the embedded object or from the corners of the mold and result in a faulty finish.

If, on the other hand, you cover the mold too late, too much styrene will have evaporated from the surface of the resin and inevitably produce a tacky film. If it is removed with solvents, sanding and polishing will be indispensable and cause extra work.

The Right Mold

The quality of the mold is obviously reflected in the quality of the casting. The manufacturers of embedding kits offer a wide range of different molds. There are fairly large differences, not only with regard to the size and shape but also to the quality of the molds.

In very cheap kits, you may sometimes find rather primitive, vacuum-formed molds with a very thin wall thickness, little stiffness and only a moderate surface quality. All this will be reproduced in the cast, the shape of which may be impaired and the finish neither clear nor glossy. Such a mold may not even stand up to repeated use, as the chemical hardening is an exothermic reaction which produces so much heat that, with the bigger castings, the mold temperature may exceed 212° F (100° C), which thin molds cannot withstand. Injection-molded polypropylene molds are, in fact, more expensive, but they are superior, first, because they are self-releasing, so that you can dispense with release agents, and second, because they have a high-gloss finish, which automatically results in crystallike casts that require neither finishing nor touching up. Their greater wall thickness also ensures sufficient dimensional stability, which is maintained if the mold gets hot during the hardening process.

Besides such glossy plastic molds, you can also use containers made from tin (which must not be lacquered), china, glass, acrylic glass, polythene, PVC and even wood, if the latter is given several coats of DD lacquer after careful sanding or, at least, treated with several coats of hard wax.

The mold must not be wider at the bottom

When choosing a container to serve as a mold, you should consider that it should not widen at the bottom, which inevitably leads to problems when removing the cast from the mold. You may even have to destroy the mold to remove the cast.

Molds from china, enameled metal and acrylic glass (Perspex, Plexiglas) are most suitable, as these materials are self-releasing with polyester resin. With glass molds, you will also have little trouble, as polyester does not usually stick to glass. Nevertheless, to be on the safe side, polish glass molds with a little release wax.

The size of the mold depends on the dimensions of the object to be embedded. As a rule, it should always be encapsulated by a coat of resin that is ½-in. (1-cm.) thick all around. Once you have selected the matching mold, you should wash it with lukewarm water to which a small amount of a wetting agent (e.g., a granulated detergent) has been added. This is to remove dust and dirt particles. You can also rinse the mold with solvents and dry it with a soft absorbent rag, which must not, however, leave any fluff in the mold. Molds that are cleaned with water and detergent can be dried with a similar cloth in a dust-free place.

Measure the capacity when cleaning the mold!

When cleaning a mold, you can, at the same time, check its capacity. In purpose-made molds, this is usually shown on the base.

For measuring the capacity of a mold, you should fill it with water and weigh it. Deduct the empty weight of the mold from the measured weight with water. The figure worked out by this method is the basis for the following calculation: as the specific gravity of polyester resin is about 10 percent higher than that of water and as 5 percent to 10 percent of resin is left in the mixing pot, you should add 15 percent to 20 percent to the measured weight of the water in the mold in order to find the required quantity of resin to be mixed with hardener.

The Casting Process

Cast only thin layers!

About one-third of the calculated quantity is mixed with the proportionate quantity of hardener and cast into the carefully cleaned and well-dried mold to form the basic layer. After about fifty minutes, this coat of resin will have gelled, so that you can now glue the object you want to embed on the surface of this layer with a little resin mixed with hardener. This is to prevent the object from floating or being dislodged when the next coat of resin is poured.

To prevent bubbles and cracks due to heat accumulation in the hardening resin, which has poor conductivity, it is advisable not to complete bigger castings by casting only one thick finishing coat of resin on top but to build the block up from several thin layers cast one by one. Allow each layer to harden and to cool down completely before the next one is cast on top.

Thin layers reduce the risk of air bubbles, as the air that is entrapped in the pores of the object to be embedded will not have so far to travel to the surface of the still liquid resin. Above all, smaller mixes are better because they produce less reactive heat and harden less quickly. Thus, the air particles sticking to the surface of the object do not expand so much and even have more time to rise to the surface of the resin. When casting the last layer, you may be a bit more generous and allow the level of the resin to rise above the upper edge of the mold by approximately 1/32 in. (½ mm.), which is possible due to the surface tension of the resin. As long as the resin is still liquid enough to allow air bubbles to rise to the surface, where they burst, you should by no means cover the surface. Here, the change of color is a great help in deciding just when to cover the resin.

The last coat may be a bit more generous

Cover the resin at the right moment

As long as the resin is still liquid, you may even guide air bubbles to the surface with a not-too-thick needle and, if necessary, make them burst. But as soon as the gelling has started (change of color), you must refrain from such actions, as the resin is now no longer liquid enough to fill invisibly the channel caused by the needle. When covering the mold, you can use a trick to prevent air getting in between the surface of the resin and the cover. If a self-releasing plastic sheet (Melinex, Mylar, Hostaphan) is used, we recommend that you roll it on a pencil or a length of doweling and then unroll it onto the resin from one edge of the mold, like an awning. If, however, you prefer a piece of glass as a cover, you should slide it over the mold from one side, with the glass lying tightly on the edges of the mold.

Allow the covered mold to cure at room temperature for some hours and wait for the shrinking of the resin, which accompanies the hardening and makes it much easier to remove the cast.

Once hardened, the cast will quite often drop out of the mold when turned upside down. If this does not happen at once, it may well do so a little while later. Impatient people may simply press a hook with a sucker, as used in the kitchen on ceramic tiles, on to the block and pull it out of the mold without damaging either cast or mold.

The Finishing Touch

If it turns out that the cast has a small flash at its upper edge after being removed from its mold or if there nevertheless happen to be some small bubbles on the previously covered surface that impair the finish of the cast, they can be removed by sanding with wet and dry paper on a sanding block.

Start with fairly coarse wet and dry paper and then change

to more and more finely grained paper. But always make sure that any marks caused by the coarser paper used before have completely disappeared before you take the next finer grade of wet and dry paper.

Extremely good results can be achieved if you take the last few courses of sanding with very fine paper and water, to which you may add some soap or wetting agent.

Special polishing paste will produce a final super-gloss finish. Do not use just any polishing paste. The best results are achieved with special polishing compounds for polyester, which are adapted to the hardness of polyester resin and may even increase the brilliant glossy finish of the surfaces produced automatically by a good mold.

Special polishing compounds for polyester

In addition to the standard use described above, the technique of embedding also opens a wide scope of attractive variations. You may tint casting resin to transparent or even opaque shades of color by adding special color pastes and thus achieve interesting effects and contrasts. Thus, for instance, you may put a bright yellow everlasting flower on an opaque blue-tinted basic layer of resin and then finish the block with several layers of clear resin.

How to Dry Flowers

On the subject of flowers, it is necessary to bear in mind that due to their moisture content, fresh flowers cannot be embedded. You have to dry them first—otherwise the resin will turn milky.

Fresh flowers cannot be embedded

Drying is best done in a wooden container filled with dry dust-free sand, such as that used in bird cages. Due to its fineness, it is highly water absorbent. Carefully embed the blossom in the sand and put the little container on a heater, where it is left until the flower is completely dry. Then, you must gently remove the sand with a very soft brush, readjust the petals of the flower and apply two coats of a good-quality hair spray. The flower is then ready for embedding.

You can also dry flowers in a can filled with dry washed sand (e.g., sand for fish tanks), to which about 10 percent of silica gel has been added. Silica gel is a strongly water absorbent chemical, which can be bought in a drugstore or a specialist shop. It is not wasted when it is used for drying flowers, because it can be reactivated in a baking oven again and again by heating it to a temperature of 248° F to 302° F (120° C to 150° C).

Having absorbed water, the silica gel turns more and more rose-colored, which shows that its water-absorbing capacity is nearly exhausted. In a completely dry state, silica gel has a bright blue color. The best way of drying flowers is to place them upright in a tall container, which is then carefully filled with the drying mixture of sand and silica gel. Once the flowers are completely buried, seal the container with a tightly fitting lid, which can be additionally sealed with masking tape. Allow to dry for about three days. Then pour the sand out of the container and clean the flowers, which may now be embedded straightaway. This method has proved most effective for most flowers, especially for roses and other red flowers.

Besides silica gel, you can also buy special salts that promote the drying of flowers.

Desiccators, which are frequently used in chemical laboratories for drying, cannot be used for preparing flowers, as they contain sulphuric acid as a water-absorbing medium, which is most dangerous for the amateur and also causes the colors of flowers to fade.

Additives and Amber Resin

Besides transparent or opaque tinting of the resin, you can also add amber resin, which is activated with paste, to the clear embedding resin. By adding 10 percent to 30 percent of amber resin to the clear resin, you achieve a light to deep amberlike shade. Pendants cast from such a bright yellow mixture can hardly be distinguished from genuine amber, if you bring some whitish spots into the casting, as they are quite often found in real amber, too. These spots are made from a small quantity of resin mixed with a little white color paste. Apply this mix with a small brush on the first coat of resin cast into the mold.

Amber effect

By adding some aerosil powder, you can make the whitish resin a bit pasty, so that you may even mold it in any desired shape and prevent it from running. Thus, you can achieve a very natural-looking imitation of amber.

Fissure effect

Finally, you can also enliven the polyester cast by adding a crack-effect paste, which is offered by one manufacturer of embedding kits. Only 0.2 percent to 2 percent of this paste produces an attractive filigree work of smaller and bigger cracks inside the block, which, however, do not reach the surface of the block and so increase the impressive effect. Unfortunately, the crack-effect paste is highly toxic and needs to be kept away from children.

The use of crack-effect paste combined with amber resin is most attractive. The cracks result from the release of a gas from the crack-effect paste due to the exothermic hardening reaction. The gas is released at a time when the gelling of the resin has already reached an advanced state, which does not allow the formation of bubbles but makes the gelling resin crack instead. By varying the amount of paste, you can control the formation of cracks. As the heat of reaction plays an important role in this process, less crack-effect paste will be required when casting large blocks than with smaller ones to achieve the same effect (due to the accumulation of heat in large blocks).

For the same reason, you will need less hardener for large casts (normally 1.5 percent to 2 percent). When making a mix of 7 oz. to 35 oz. (200 g. to 1,000 g.), only 1 percent will do; with mixes between 2.5 lb. and 22 lb. (1 kg. to 10 kg.), even 0.8 percent is enough. More than 3 percent of hardener will turn clear casting resins slightly yellowish. If you use cobalt accelerator, you will have to put up with a slight bluish-violet shade of the hardened resin. Generally, you can dispose with additional accelerator, especially since slower hardening is an advantage, as it involves less risk of heat accumulation and unwanted cracks. With very large casts, it may become necessary to cool the block while curing, to carry off the heat of reaction and to keep the risk of tension cracks as low as possible, even with the lowest possible quantity of hardener and no additional accelerator. This applies especially to mixes of several pounds of resin. *Hardener/accelerator*

Embedding offers an enormous scope, allowing free range to imagination and talent. The number of objects that can be made from casting resin seems to be limitless. Lamp bases, book ends or rests, pendants, decorative knobs and handles for a corkscrew, for the utensils in your home bar or for the doors and drawers of a cupboard, decorative panels that may even be illuminated from the back, pen trays, small bowls and dishes for the party table, ash trays, name plates of any size, pieces for games and many other items can be molded from casting resin in any desired size and shape. *Wide scope*

Elements such as jewelers' findings can be incorporated in the cast, and these will allow any fittings to be secured by screwing or bolting. Avoid point loads, however, and try to disperse any loads in a way appropriate for the material involved.

There still is one important point this book cannot influence nor is intended to influence and that is the problem of the very versatility of the resin, which may lead to dubious experiments.

Bear in mind that a good job must do justice both to the material and the practical use of the object.

CASTING RESIN FOR THE ELECTRONICS ENTHUSIAST

Epoxy resins for electronic circuits

In the field of industrial electronics, it has been standard practice for years to protect delicate circuits against mechanical stress by embedding them in plastic resin. The amateur can adopt this technique. Their shrink resistance makes epoxy resins especially suitable for potting electronic circuits, as the resin prevents any strain on the electronic elements and their wiring. Nevertheless, you can also use crystal clear polyester casting resins. It is most important with all work of this kind to avoid, as far as possible, the generation of excessive heat, which is detrimental to transistors and other semiconductor elements. The best way of doing so is to use less-reactive types of resin and to cast the block in several thin layers, in order to avoid heat accumulation. When planning a potted circuit, you must, however, consider one fact: once potted in resin, a circuit is very difficult to repair. The only way of repairing it at all is to cut the defective parts out of the block with a saw and try to bare the connections in order to be able to solder the spare parts into place and re-embed them in resin.

Electronic minicircuits
in module block

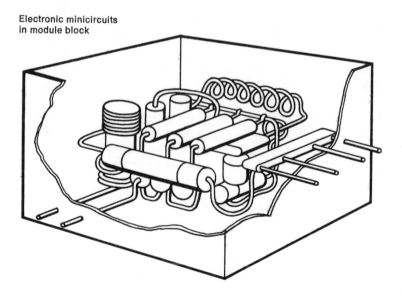

Industrial firms make use of the so-called module system, where single parts of a circuit are potted as integral electronic minicircuits, which can then be combined with plug contacts to make a complete circuit. In case of defect, the faulty module can be replaced as a complete unit. In order to avoid problems due to poor contact or bad fittings of the plugs, this technique requires a high degree of precision when embedding the components, so that this field is more or less the preserve of experts in both electronics and plastic resins.

ENAMELING WITHOUT A KILN

Epoxy resins that harden with a nontacky surface can also be used by clever amateurs for cold enameling, for which resins tinted with bright color pastes are used. The decorative effect of cold-enameled objects is by no means inferior to genuine enamel work fired in a kiln. Modern types of resin and color pastes give them the same bright and brilliant colors. Only the hardness of cold enamel is not so great as that of genuine enamel. Cold enamel is less subject to chipping, but resistance to weathering and light, however, are not yet completely satisfactory. However, development is continuing, and though a number of cold enameling sets have been on the market for quite a long time, you may count upon further improvements in the future. Some manufacturers, for instance, try to reduce the low scratch resistance of cold enameling resins by adding a kind of lubricant to the resin. This settles on the surface of the resin and forms a film, which reduces the frictional resistance and, thus, reduces the risk of scratches in the finished product.

Furnace enamel differs only in hardness

All the epoxy-based cold enameling resins on the market have a fairly long curing time. As a rule, such resins require approximately twenty-four hours to harden, even though some products stop being tacky after three to four hours. When exposed to pressure, they show impressions that indicate that the resin is not completely hardened. Depending on the type of resin and the quantity of the mix, the pot life varies between twenty and forty minutes. Some resins must be allowed to prereact before they are applied, i.e., you must allow them to stand for a while after adding the hardener.

Pot life twenty to forty minutes

Cold enameling resins are most interesting wherever kiln enameling is impossible, because the object would not withstand the heat or is too big to fit into a kiln. You can apply cold enamel to trays and tabletops made from wood, chipboard or laminated glass and to ceramic or china objects. But you should try to avoid direct contact of cold enamel with food. A cold enameled top for a coffee table made from artificial stone is most attractive, too, if you use cold cast enamel objects instead of a polyester gelcoat in a box-type mold. Once the enameling resin has started to gel, a thin coat of dry sand is spread on top and allowed to harden thoroughly. Then, any loose sand particles are brushed off before a thick coat of resin-bound mortar, mixed from dry sand and polyester, epoxy or one-component PU resin is applied, as described on pages 111–112. Solidify this coat by rolling it with a length of round wood, apply reinforcing wire and fill the mold with another coat of resin mortar. Thus, you obtain a solid table-top with an entirely smooth surface, embellished with cold enamel ornaments.

As a matter of fact, you can also use cold-coat enamel to cover a normal tabletop made from stone, artificial stone or wood, but this will not have an absolutely smooth surface unless you work most carefully. Owing to the very good self-leveling characteristics of cold enameling resins, the surface should never-theless turn out to be smooth enough to allow glasses and bottles to stand safely on it.

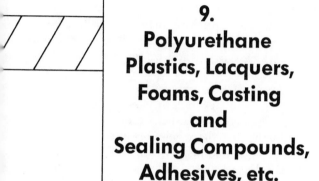

9.
Polyurethane Plastics, Lacquers, Foams, Casting and Sealing Compounds, Adhesives, etc.

The history of this most interesting group of plastics began in the summer of 1937 in the research department of the Bayer company, which at that time belonged to the IG-Farben trust.

In those days, Otto Bayer charged his chemists with the task of making di-isocyanates and trying to compose macromolecules from them. This task must have appeared to the chemists in the research department as a dream, just as squaring the circle must appear to a mathematician.

Monoisocyanates had been known for quite a long time. They are characterized by a highly reactive group consisting of one atom of nitrogen linked to one atom each of carbon and oxygen, with a double bond between the carbon atom and its neighboring nitrogen and oxygen atoms. This double bonding is the reason why this group is so reactive. As a single bond would be quite enough to link the atoms, chemical substances with double bonds tend to transform this double bond into a

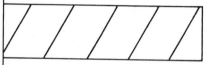

Molecular structure

single bond, and then use the now free additional bond to combine with other reactive molecular groups.

It was known to Bayer that the monoisocyanates mentioned above readily combined with alcohol by simple addition to form new combinations, called urethanes. He now started with the idea that a di-isocyanate—a combination that contains two highly reactive $-N=C=O-$ groups instead of the one contained in monoisocyanates must provide a chance for a double bonding, allowing the formation of extended molecular chains or even netlike molecules. Much was said in favor of this theory, but there were also less favorable omens. Quite a considerable technical expenditure was required to produce monoisocyanates, which made them costly. Above all, it was very difficult to control these reactive substances, which had already caused some headaches to the chemists in their laboratories. And now Bayer was asking his chemists for nothing less than di-isocyanates.

The origin
of a new
family
of plastics
Gritting their teeth, the Leverkusen chemists began to investigate isocyanate compounds, and many a man in the laboratories must have thought that the whole job was a sheer waste of time. But things took an unexpected turn. Only four months after the research into di-isocyanates had begun, the chemists in Leverkusen not only managed to produce the di-isocyanate but also succeeded in creating a new family of plastics by a controlled reaction between di-isocyanate and polyhydric alcohols, which include glycerine. This was the birth of the first PU resins.

It soon proved that this trail blazed by Bayer and his team opened a wide scope of most interesting applications, as the properties of this new plastic could be influenced and controlled in a way nobody had dreamed of with any other plastic. The chemists not only succeeded in making meltable chainlike polyurethanes for the production of fibers but also managed to construct nonmelting interlaced macromolecules. By a clever choice of the interlacing partner (polyhydric alcohols) and by controlling the proportion of di-isocyanates serving as a hardener and of the alcohol groups, the hardness of the interlaced types of this new plastic could be adjusted at will between soft resilience and glasslike brittleness. Thus, the chemists were able to produce tailor-made synthetics for many applications.

Quite soon, the chemists also learned to make use of the tendency of di-isocyanate to disintegrate in the presence of water and to split off carbon dioxide, as baking powder does when it is heated. This could be used to expand PU plastics by a controlled reaction of a precisely proportioned quantity of water with a proportionate quantity of excess di-isocyanate, which is not required for interlacing but causes the resin to foam instead.

In the following years, the development of PU resins was pursued with much energy, but, for quite a long time, PU materials remained rather tricky and fairly expensive substances.

Do-it-yourself Potential

Since then, the extraordinary qualities of the polyurethanes have earned them a major position among the other synthetics on the market, and they are still gaining in importance. In some cases, they have become superior to other types of plastics. New methods of production and the transition from small-scale to mass production reduced the price barrier, so that PU resins are readily attainable today. Even their application has become easier, and amateurs have been able to use them for some years. PU resins are offered in great variety as lacquers (so-called DD lacquers), foams, coating and casting compounds, adhesives and sealing materials.

Above all, the popular leatherlike material, which is both decorative and easy to clean and also allows air to get through while water is rejected, is based on a PU product applied to a suitable fabric. Liquid PU resins are of special interest to hobbyists and craftsmen, and their special properties are extending their application. PU resins stick naturally to virtually any material. Even absolutely smooth and completely nonporous surfaces are not much of a problem.

The secret of this astonishingly good adhesion is the fact that, contrary to many other plastics, polyurethanes do not have a measurable shrinkage. The large scale of different formulas between soft and resilient and brittle rigid types also proves an advantage. Moreover, polyurethanes have excellent electrical properties, a good resistance to chemicals and they are extremely weatherproof.

Disadvantages and Possible Dangers

Nevertheless, there also are some negative aspects. Many PU resins have a tendency to turn more or less yellowish after some time, and they are susceptible to water as long as they are liquid. Moisture may easily cause bubbles or turn the solid material into foam, which is the result of the already mentioned tendency of the di-isocyanate to split off gaseous carbon dioxide due to its reaction with water.

Toxic hardener

A further negative aspect is that the di-isocyanate components that serve as hardeners are highly toxic. Therefore, much care and precaution must be taken. Especially when PU resins are applied with a spray gun, a breathing mask with fresh air supply is indispensable. Amateurs who rarely have such equipment should, therefore, refrain from applying polyurethanes with a spray gun. Due to the toxicity of still liquid polyurethanes, you should carefully observe the safety precautions of the manufacturers. The most important rule is to ensure good ventilation of the work place.

Assuming a proper proportion of the basic component and the hardener and proper mixing, polyurethanes, once turned hard, are no longer toxic and are physiologically safe.

Limited shelf life

The limited shelf life is another problem presented by polyurethanes. In tightly sealed containers, most PU systems can be kept for about six months, if they are stored in a cool or only moderately warm, dry place (40° F to 68° F [5° C to 20° C]). Once opened, the containers cannot be stored for such a long time. Some PU foams and colored one-component materials have an even shorter shelf life.

During protracted storage, the interlacing component (hardener) may possibly produce a crystallized deposit, which must be dissolved by thorough stirring before the hardener is used.

PU resins are supplied in two different basic types: as one-component systems and as two-component resin hardener systems.

ONE-COMPONENT POLYURETHANES

Lacquerlike or pasty

One-component PU plastics are, first of all, supplied as solvent-containing lacquerlike materials, which, however, have a much wider field of application than lacquers. Moreover, there also are pasty one-component PU compounds that are used as filler. Both systems have a similar hardening mechanism. As there is no reactive partner in the shape of a second component, only so-called prepolymer resins are used for such purposes. The structure of these resins can be compared with prefabricated elements to be combined into an assembly kit. The liquid consists of pre-interlaced molecular groups. The interlacing is, however, not yet advanced to a state where the liquid resin changes from its original liquid state into a solid state.

The prefabricated molecular elements or fragments of the final molecule only combine to form a macromolecule by a chemical reaction if they are activated from outside. This process is initiated by air humidity but by no means by moisture, which will inevitably lead to the formation of foam and bubbles.

Solvent-containing Liquid Systems

The hardening of lacquerlike one-component PU resins is blocked by solvents and by the exclusion of air, as long as the resin is kept in its container. As soon as you allow the solvents to volatilize and humidity from the air gets to the resin, hardening will begin. If you want to ensure proper curing, take care that lacquerlike resins are only applied in fairly thin coats, to prevent the resin from hardening too quickly at the surface while the material underneath is still liquid and cannot get rid of the solvents. Too thick application results in bubbles, as well as poor curing, as the vapors of the volatilizing solvents will try to penetrate the surface, which is sealed by a thin, hardened coat of resin. The condition of applying only coats with a certain maximum thickness not only applies to the use of PU resins as lacquers or varnishes but also to their use as primers, especially when the resin is used for plastic mortar.

Good hardening in thin layers

Before dealing with some examples of uses, we should have a look at the most important characteristics and safety precautions:

PU ONE-COMPONENT LIQUID RESINS IN PRACTICAL USE

APPLICATIONS: sealing of wood, concrete, brickwork, corrosion proofing of metal, turning slippery floors in wet rooms and stairs into nonskid, primers for surfaces to be coated with polyester or PU resins, insulating wet cellars and walls, repairing concrete floors and worn stairs, making artificial stone plates.

CHARACTERISTICS: yellowish to brownish color, similar to thin tea or even strong coffee in the case of highly interlaced liquid resins. Strong smell of solvents.

CAUTION: avoid fire and ensure good ventilation because of solvent content. Splashes onto mucous membranes or into the eyes are dangerous. The resin will leave brown stains on the skin if splashes are not removed immediately. Wear protective polythene gloves or at least treat your hands with barrier cream before you start working. Once turned hard on your skin, the resin can only be removed with pumice stone, which may take the skin off as well. It is probably better to wait until the stain grows out—when normal regeneration causes shedding of the top layers. Liquid one-component polyurethanes are toxic (neutralize immediately if they get into contact with your skin or wash away with solvents; if any of it gets into your eyes or if accidentally swallowed, call a doctor immediately). Never burn spilled resin, because highly toxic cyanogen compounds will be set free. Remove spilled material by putting sawdust or, in emergencies, dry sand on top and neutralize with plenty of clear water.

Insulating a Cellar with Liquid Plastic Resin

In view of increasing leisure time, cellars become more and more useful. But quite a few people who planned a "hobbyist's paradise" have returned to the light of day bereft of their illusions. They had indeed expected to find a dark hole—cellars are dark by nature—but what they had found was enough to daunt the greatest optimist. The stairs down to the cellar may have proven to be a dangerous trap, with their worn steps; the floor of the cellar may have been cracked or riddled with holes. Crumbling cellar floors are a never-ending source of dust, which covers everything with a dull coat of gray cement. Finally, there might also have been a wet wall in the cellar, which makes the hobbyist imagine how his beloved and well-cared-for tools would look after being kept in this damp room for any length of time.

But even if all this trouble besets just one poor mortal, he need not abandon his plan of using his cellar as a den. All these problems can be solved with one big can of a one-component polyurethane.

How to Dry Wet Walls

The widespread trouble of wet cellar floors and walls can be easily banished with a liquid plastic resin which penetrates into the pores of concrete or brickwork and there forms a tight sealing zone. As the insulation does not take place on the very surface but inside the stone or concrete, it cannot chip off.

In order to achieve the deepest possible anchorage, the resin should be brushed into the wall or floor with a stiff-bristled brush. Because of the risk of splashes, we suggest you wear goggles. Depending on the porosity and absorptive capacity of the ground, a single coat will often do, or two or three may be required. In case the wall or floor to be treated is very dense and only has very fine pores, it may be advisable to add thinners to the one-component material in order to allow it to penetrate more deeply into the base.

Normally, damp surfaces that are cool and slightly clammy to the touch can be sealed without prior treatment, providing loose particles of lime and possible efflorescence have been removed. Wet surfaces must be dried before the sealing agent is applied, in order to avoid the formation of foam. In certain circumstances, it will be quite sufficient if the room is ventilated for a couple of days, provided that the weather is warm and dry. If this does not help, you will have to bring a bigger gun into action and try to dry the room by heating or by treating the wet surfaces with a propane torch.

Dry wet surfaces before sealing them

However, you must not work on one side with a propane torch or another kind of heater while the opposite wall is being treated with a solvent containing PU solution, which releases large quantities of inflammable vapors during the drying process.

If several rooms are sealed one by one, you should allow sufficient time for the solvents to volatilize completely before the torch is used again. In order to reduce the effect of solvent vapors, ensure good ventilation when the sealing agent is applied. A slightly glossy superficial film indicates successful sealing. Apply coat by coat until this film is achieved. Coats are always applied at intervals of three or four hours, to ensure proper adhesion between coats. Once sealed, the walls can be given a coat with PVA wall paint, but one can also tint the sealing agent with special pigments, if darker shades of color are chosen—which must be recommended, anyway, because of the brownish shade of the resin. You must, however, be prepared to accept a slight yellowing if the sealing is exposed to ultraviolet light.

Bad Floors Repaired with Ease

Just as walls can be efficiently insulated against penetrating moisture, you may also insulate concrete or cement floors. By treatment with a liquid plastic resin, old, dusty floors will acquire a smooth abrasion-proof surface, which is easily cleaned.

Before the sealant, the second coat of which may also be tinted, is applied, the floor must be thoroughly cleaned with a vacuum cleaner. Wherever you find cracks or even holes, you will have to fill them in before sealing the floor.

Applying
plastic
mortar

This is best done with a plastic mortar, which you can mix yourself from dry fine sand and liquid resin. The best mixing formula is 85 percent to 90 percent sand plus 10 percent to 15 percent resin. To reduce the risk of bubbles, you can dry the sand in a domestic oven, but allow it to cool down again before mixing with liquid resin.

Damaged parts of the floor, such as cracks and holes, have to be cleaned and any loose parts removed and widened, if necessary, to allow them to be filled in with repair mortar.

In order to ensure a good adhesion of the mortar, the holes and cracks should be given a priming coat of the liquid resin, which is allowed to gel until its surface is only slightly tacky. Mix the dry sand with resin in a PVC or polythene bucket. If you observe the mixing proportion given above, the plastic mortar will have a consistency similar to that of the normal dry to stiff consistency of cement mortar.

Mind the
grain!

It is most important that the proper grain of the sand is observed. This is necessary to make sure that the solvents contained in the resin can volatilize properly, which is a prerequisite condition for perfect hardening. Layers of a maximum thickness of 25/64 in. (approximately 1 cm.) are advisable for the same reason. Deeper holes must be filled step by step in several layers. As the mortar can only be applied within twenty minutes, you must not mix too much mortar at once.

Short
pot life

The fresh mix of plastic mortar is applied to the cracks and holes previously treated with primer with a slender tapering trowel or with a spatula and is then smoothed with a smoothing trowel wetted with solvent.

If you do not intend to give the whole floor a coat of liquid plastic resin, the repaired surfaces receive a finishing coat of sealant, which is applied about three hours after the mortar, but if you want to seal the whole area, the sealing of the repaired sections is done when the whole floor is sealed.

Roughening
of the
floor

Concrete floors treated with cement slurry must be roughened mechanically or chemically in order to ensure a proper adhesion of the sealing coat.

Without special equipment, mechanical roughening is extremely tiresome, so that most amateurs prefer the chemical way and treat the floor with fuming hydrochloric acid, which is sprinkled onto the floor with a plastic watering can before it is dispersed with a scrubber. Then, rinse with plenty of water and allow the floor to dry for at least one day.

If you want to apply a colored sealant, it is advisable to apply the first coat without adding pigment, in order to keep the risk of bubbles as low as possible. You may also give the floor a first priming coat with liquid resin and then apply a colored sealing agent that is based on the same type of resin and can be bought in different colors.

Oil Stains and Slippery Floors

As one-component PU sealants have excellent resistance to oil and chemicals, they are most suitable for sealing floors against these media. Therefore, it is a good idea to seal the floor of a garage so that you can forget the unpleasant oil stains on the floor forever. Existing stains must, however, be removed with hot alkaline cleaner or a special oil-removing agent before sealing, to ensure good adhesion.

Good for garage floors

If your garage is big enough so that it is not only used for parking your car but also for maintenance work and washing the car, you will surely remember without much enthusiasm how slippery the garage floor can become when wet. In such a case, the sealant is a reliable remedy, also, if you scatter dry arenaceous quartz sand or carborundum granules over the sealing compound while it is still liquid. Sweep away any loose particles after three hours and apply another coat of the liquid resin for final sealing. This will stop the risk of skidding, which is a major cause of accidents.

This trick works not only on concrete or cement floors (garages, terraces, stairs in the garden or outdoor stairs leading to a cellar, washhouses, etc.) but also on metal surfaces (stairs, boat decks, covers for fuel tanks, etc.).

As a matter of fact, it is a condition for good adhesion that the surfaces to be treated with PU sealing agents are well prepared and thoroughly free from dirt, grease and rust.

When painting iron or steel, great care must be taken to remove grease and flaky rust completely. Fine flue rust is no obstacle, as it is bound by the resin. Because of its good resistance to weathering and abrasion, this material provides excellent protection against corrosion.

Painting of Tropical Woods—No Problem

One-component PU resins can be used to paint tropical wood and provide a lasting protective film. This material is especially suitable for sealing tropical woods containing special substances that quite often inhibit the drying of conventional lacquers.

For light-colored woods—a slightly yellowish resin

It must, however, be remembered that the brownish color of the resin makes wood appear darker, which might be advantageous with some types of wood but undesirable with others. Therefore, it is advisable to use the only slightly yellowish type for sealing light-colored wood, even though this type of resin is slightly less resistant than the brownish types of PU resin, with their higher degree of interlacing.

Even here, an interval of one to four hours between coats must be observed if several coats are needed. Otherwise, you will have to sand one coat before the next one can be applied.

Primer

We have already dealt extensively with the most important role of one-component PU resins, as primers for glass fiber coatings. In addition, they also serve as a reliable primer for two-component PU resins. They seal the ground against moisture rising from underneath, which may lead to the formation of bubbles, as two-component polyurethanes are fairly susceptible to water. Above all, the primer blocks the air-containing pores in the ground, from which air might otherwise rise due to the expansion caused by the heat of the hardening reaction and cause bubbles.

Priming coat

At the same time, the very fluid primer will penetrate deeply into the underground and provide a good anchorage for subsequent coats. Due to the close chemical relationship between the primer and the final coating, the two-component resin must be applied within one to four hours after the primer to ensure the best possible adhesion.

What Else Can One-component PU Sealing Agents Do?

For the keen do-it-yourself man who likes experimenting, this material proves to be a boon in many jobs that used to be problems. Unfortunately, this book is too short to enumerate all the uses, but here are a few hints.

One-component PU resins can be used for:

Sealing, binding, impregnating

- sealing porous ceramic jars or vases by only rinsing them with resin;
- the gravel surfacing of roofs can be bound by simply spraying resin over the pebbles;
- molds made from wood or plaster, which are intended for laminating or casting purposes, are perfectly sealed with resin.

The sealing coat must be applied not less than four hours before the mold is used. Otherwise, the resin will act as a primer and obstruct the removal of the cast. According to the formula for plastic mortar given above, you can also make artificial stone tiles, using dry sand with a grain-size distribution of 1/128 in. to ⅛ in. (0.2 mm. to 3.0 mm.).

EXPANSION JOINTS

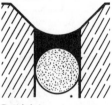

Butt joint

Overlapping joint

• Foam profile limiting the depth of the joint to be filled

PU FILLING COMPOUNDS AND THE FILLING OF JOINTS

Many a do-it-yourself decorator has all but despaired of the rather trivial problem of sealing a joint efficiently, and he may now nearly be inclined to regard this chapter as a sort of occult science. Bad experiences tend to make us see a problem as much bigger than it actually is, and the sealing of joints is a problem indeed—if one does not know that joints are not all of the same type and that every filling agent is not equally suitable for each and every kind of application.

Therefore, we should first briefly deal with the different kinds of joints.

Professionals distinguish among expansion joints, like those we know from large concrete areas; connecting joints, as you can find between the brickwork and the window frame; sanitary joints between bathtubs or shower stalls and the tiles of the walls and, finally, a fourth type, the so-called dummy joint, which does not actually cut the material into two parts but serves as a means of styling, in order to make large sur-

Four types of joints and how to fill them properly

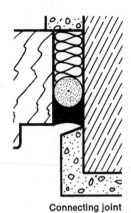

Connecting joint

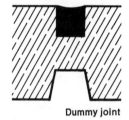

Dummy joint

faces appear less dull and monstrous by dividing them into several smaller sections. Another reason for dummy joints is the wish to reduce shrinkage stress during the hardening of concrete surfaces.

All these types of joints are exposed to different loads and stresses and, therefore, also require special jointing compounds for filling.

Expansion joints are designed to balance and take up the moving of parts of a structure caused by temperature, humidity, creeping or setting and prevent damage to the building from such movements. They must also prevent water or any kind of humidity from penetrating into the joint. The demand for flexibility precludes the use of any rigid sealing agents. If this point is not considered or if the necessary expansion joints were omitted, all the surfaces will crack. In harmless cases, such cracks can be transformed into proper expansion joints, but they may also lead to enormous damage. Therefore, one should strictly refrain from filling such joints with cement or concrete. There may also be movement between the brickwork and the frames of windows and doors, so that absolutely rigid joints between such parts are not permissible. It is necessary to use a suitable filling compound in order to seal such a typical connecting joint.

The third kind of joint, the sanitary joint, is static. The filling of such a joint is first of all done for hygienic and optical reasons. A filled sanitary joint is much easier to clean and avoids the risk of growing germs and fungi. At the same time, such a joint will look much nicer. Ever though such a joint does no work, the filling compound must resist manifold stresses, such as constant exposure to water and household cleaning agents.

Which Filling Compound for Which Purpose?

Permanently plastic compounds

Technicians distinguish two basic types of filling compounds: permanently plastic and permanently elastic compounds. Plastic compounds deform under load and do not have the property of returning to their original shape. They are pasty, like Plasticine, and generally produce a tight skin on their surface, which is replaced by underlying material if damaged. Such compounds usually contain solvents and shrink a little due to the loss of solvents through evaporation, which leads to a slightly hollow surface.

Permanently elastic compounds, however, are rubberlike and return to their original shape after temporary stress. This classification is to a great extent academic, for nearly all joint-filling compounds lie somewhere between these two extreme points. They are generally classified according to their predominating properties.

Permanently plastic compounds are based on oil, synthetic resins or synthetic rubber types (polyisobutylene, butyl rubber). Polyacrylate-based compounds are semiplastic and semi-elastic. Distinctly elastic joint-filling compounds are based on polysulphides (thiokol), silicone rubber or PU resin. They are, without exception, supplied as one- or two-component systems, the latter being generally more resistant. Permanently elastic compounds are, as a rule, used for filling joints that have to stand up to much use (expansion joints), while connecting joints subjected to lesser stress are, in most cases, filled with permanently plastic compounds.

Sanitary joints are best sealed with high-quality one-component silicone rubber (see page 169). In do-it-yourself shops, you can also buy cheaper filling compounds for this purpose, which are based on PVC/PVA and quite often tend to hardening and shrinking in the course of time.

In addition to identifying a joint and choosing the right filler, there is also the correct way of filling a joint.

How to Do the Job

The width of the joint is a most important factor. It must be wide enough to allow the filling compound, by virtue of its resilience and plasticity, to withstand movements of the joint. If a joint works, for instance, by 5/64-in. (2-mm.) maximum and if the filler compound for the joint can take up movements up to 20 percent elastically, the joint must be 25/64-in. (10-mm.) wide. In order to be on the safe side, you can make it 15/32-in. (12-mm.) wide.

The depth of the joint plays an important role, too. A 25/64-in. (10-mm.) wide joint between two prefabricated concrete elements is normally filled to a depth of 25/64-in. (10 mm.). The same depth is also recommended for joints having a width of 19/32 in. (15 mm.). For joints of 1 in. (25 mm.), the average filling depth is 19/32 in. (15 mm.) and in a width of 1 3/16 in. (30 mm.), it should be about 25/32-in. (20-mm.) deep. If the joint is deeper, special steps must be taken to prevent the filling compound from dropping into the joint. To do so, you can use polythene tubing pressed into the joint or, better still, a length of closed-cell PU foam, which can be bought by the yard in do-it-yourself shops. It can be fitted into the joint, pushing it in while slightly compressing it (see sketches on pages 135–136).

In order to achieve a reliable sealing, the joint-filling compound must adhere firmly to the sides of the joint. This is the reason why these surfaces must be cleaned with special cleaning agents before the compound is applied. Rub the surfaces carefully, and change the cloth often. Roughening is another way to achieve a good bonding.

Primers improve adhesion There are primers for most filling compounds to improve adhesion. If you cannot achieve proper sealing despite careful cleaning and sizing of the surfaces, it may be that there is a good bond between the filling compound and the base but poor adhesion between the surface of the joint and the material underneath. Such difficulties are often caused by loose paint.

Quite often, one sees joints with crumbling filling, which looks like bark. This is not due to poor quality of the filler but to bad workmanship. It only occurs with joints where two working parts form an expansion joint over a third, more or less static, member, and the filling compound was applied so that it sticks to both sides and the bore (three-side adhesion). See sketch on page 139.

If the filling compound is distended, it cannot yield to the full extent allowed by its elastic properties, because it sticks to the base of the joint. As a result, the joint will crack. This trouble can be avoided by preventing the filling from sticking to the bore of the joint by applying a release agent, or, what is much easier, by lining the base of the joint with a strip of oil-impregnated paper of the same width as the joint. See sketch on page 139.

Curing time Just as with one-component PU resins, the interlacing of one-component joint-filling compounds is affected by humidity from the air. One-component polysulphide (thiokol) and silicone rubber compounds harden in a similar way. The curing time of all three systems depends on the degree of atmospheric humidity and on the depth of the joint, i.e., the thickness of the filling. The

Adhesion to the base of the joint (three-side adhesion), filling compound cracks.

The use of a release agent on the base of the joint allows the filling compound to move.

higher the humidity, the faster the compound will cure; the deeper the joint, the longer it will take before the filler has hardened throughout. The influence of the atmospheric humidity is much more distinct with PU and polysulphide compounds than with silicone rubber.

PU compounds excel by their good resistance to abrasion and oil and even remain quite flexible at low temperatures. Some types resist temperatures of up to 194° F (90° C). They are sold in disposable cartridges and are pressed into the joints by means of a manual caulking gun. A wet spatula is used for smoothing the surface of the filling. Such compounds are generally gray or beige. The filling compound used as an example hardens at the rate of 5/128 in. (1 mm.) per day at 68° F (20° C) and a relative humidity of 50 percent. That means that a 25/64-in. (10-mm.) deep filling will take about ten days to cure right through under these conditions.

TWO-COMPONENT POLYURETHANES

Two-component polyurethanes present an extremely wide field of uses due to their great variety of formulas. They can be used for repairs and improvements about the house, for modeling and many do-it-yourself jobs, including:

APPLICATIONS: coating of floors, terraces, flat roofs, undercoats for cars, joint filling, cementing, glueing of nonporous materials on nonporous undergrounds, mold making, varnishing and painting of wood and plastics. As foam: insulation against heat, cold and noise; to make boats and other floating objects unsinkable; for packing of fragile goods; lightweight structures in the field of modeling and for parts with a smooth nonporous surface and a light foam core.

CHARACTERISTICS: Hardener fairly fluid and, with some exceptions, dark brown. Smell: not uniform and, in most cases, not very distinct.

SAFETY PRECAUTIONS: Most resin components are noninjurious, although they may cause allergic reactions in susceptible people. Hardener, however, is highly toxic. Safety precautions, in general, similar to those to be observed with one-component polyurethanes. They are not flammable except for resins that contain solvents. Do not burn spilled hardener, but absorb it with sawdust and neutralize it with water.

Containers may be pressurized. Keep openings well away from your own face and the faces of other people when unscrewing caps or lifting lids. Ensure good ventilation when working with large quantities. This applies especially to prepolymer foams, which can, as a rule, be distinguished from other types by their white color.

SPECIAL WORKING INSTRUCTIONS: Adhere as closely as possible to the recommended proportions. Avoid contaminating moisture or water, which would result in bubbles or foam.

COATINGS—AN INTERESTING USE OF
TWO-COMPONENT POLYURETHANES

PU resins with a greater or lesser amount of filler can be used for abrasion-proof well-adhering coatings of concrete, metal, wood, asbestos-cement and many other materials. It is very resistant to mechanical and chemical attack. There are many special types of coating materials, the elasticity, hardness and viscosity of which make them suitable for a number of different purposes.

Polyurethanes also cure at low temperatures

Unlike polyester and epoxy resins, polyurethanes also harden at low temperatures. Curing is delayed but not blocked and will eventually be complete. The hardening is a continuous process, which can be seen in the gradual thickening of the resin.

The ideal working temperature lies between 41° F to 86° F (5° C to 30° C), but even at temperatures below 32° F (0° C), there will be no problems at all (only avoid condensation water and white frost!).

Any desired thickness of coating can be achieved in only one application when using two-component systems. The coating is tight and porous.

Special care must be taken in measuring the hardener. To make sure that the coating reaches the resistance and durability indicated by the manufacturers, mixing instructions must be followed to the letter. There is only a very small margin in the proportion of resin and hardener. A deviation of 1 percent or 2 percent is the permissible maximum.

Proper mixing is important, too. The degree of mixing can be easily judged, thanks to the dark color of the hardener. With heavily filled and, thus, very viscous resins, it is virtually impossible to activate large quantities by hand. In such cases, mechanical mixing with a powerful electric drill fitted with a mixing attachment will be found indispensable.

A Seamless Floor

Terraces, balconies, workshops, cellars and even kitchens and bathrooms can be given highly resistant and easy-to-clean floors cast in place from PU coating resin.

The floor is cleaned before coating. If cracks and holes have to be made good, this should be done in the way described on pages 132–133. Porous floors, such as cement floors and stone/wood floors, which bind moisture or entrap air, should be sized with a one-component PU sealing agent to exclude the risk of bubbles.

After thoroughly mixing with hardener and resin and adding PU pigments, if required, for tinting the compounds (which are available in several opaque basic colors), the mixture is applied to the prepared floor. At a room temperature of 68° F (20° C), the flooring material remains liquid for about one-half hour. It is spread on the floor and evenly dispersed with a comb-like spatula. Depending on the inclination of the spatula, which can be held either vertical or at a more or less acute angle, the coating will be thicker or thinner.

Flooring compounds are generally self-dispersing, so that the little corrugations resulting from the use of the comb spatula will soon disappear automatically. The smoothness will be spoiled only if one tries to disperse the material after the gelling has started. It may happen that the viscosity of the resin is already so high that the marks of the spatula cannot disperse and will remain visible.

There are flooring materials that produce an entirely smooth surface and others that produce a fine-grained finish similar to that of repoussé enamel or to the skin of an orange. Such a finish results from a small amount of wood oil added to the resin by the manufacturer and is quite useful for wet rooms, as it provides good protection against slipping.

If a thicker coat is desired, one should take a more viscous type of resin, while more fluid compounds are more suitable for thin coatings.

Sometimes it may also be helpful to coat even vertical surfaces, such as the lower parts of walls in a fuel oil store or the vertical parts of stairs, etc. Such jobs call for thixotropic resin, which will not run off or produce "tear drops."

The Old Tar-board Roof Becomes Watertight—Forever

Garages, summer houses, sheds and even verandas may have flat roofs insulated with a layer of tar board. Even though tar board proves a fairly durable and lasting roofing material, the roof will become leaky one day. Mending the roof with patches of tar board and a hot bituminous glue requires some skill and is a fiddly job. Other do-it-yourself repairs are often short-lived and may increase the damage. In the end, the entire roof may have to be replaced.

PU coatings can solve the problem and make the roof watertight. The method is reliable and simple and has been adopted by professional roofing contractors. The earlier you take steps to repair your roof, the simpler and cheaper the job, for the older and the more scarred the tar board is, the more resin is required to repair it.

The resin is used in the most simple way conceivable, as the homogeneous mixture of resin and hardener is only rolled onto the dry tar board (which must have been thoroughly cleaned). In former times, people tended to roll on only a thin coat of resin, which can take up any possible movements of the underground by its own elasticity. Today, more and more often, a reinforcing layer of fabric or scrim is laid onto the still wet first coat of resin, which has been applied with a sheepskin roller. It is then covered with another coat of resin rolled on in order to embed the fabric completely.

In most cases, a rather coarse-meshed polyester fabric with a mesh size of 3/64 in. (approximately 1 mm.) is used. Roofs treated this way will be waterproof virtually forever and require no maintenance. Unfortunately, a slight yellowing of lighter colored resins, resulting from ultraviolet rays, is unavoidable.

By the way, for sloping roofs, a thixotropic type of resin should be used, as it does not run off and ensures even thickness.

A Second Skin for the Bottom of Your Car

The very good adhesion of PU resins to metal opens a new way to provide a car with a durable plastic skin that protects the car against corrosion from underneath. Its elasticity even resists the bombardment of stones hitting the car on the road, and neither salt nor oil, grease or fuel can attack such a coating.

It is, however, a prerequisite condition for a good adhesion of the coating that the bottom of the car be clean and free of any residue of oil or grease. Coating materials suitable for this purpose may be applied either by brushing or spraying.

For parts subjected to special stress, it is advisable to apply several coats of resin in order to achieve a sufficient thickness. Coating the bottom of a brand new car is the simplest way, as the underside is still clean or can be easily cleaned. It is self-evident that the exhaust system and the lubrication nipples must be kept free of resin.

Grouting with Two-Component Compounds

Besides the one-component joint-filling compounds already mentioned, there also are joint-filling compounds that are mixed from two components. They are either pourable or are squeezed into the joint by means of a polythene bag similar to those used by housewives and pastry cooks for decorating cakes. If need be, such a squeeze bag can even be improvised by filling a thick polythene bag with the mixture of ready-to-mix resin and hardener and cutting off a triangular section at one edge after the bag has been tightly sealed at its top end with a length of string. Pourable, self-dispersing compounds are the right choice for horizontal joints. Vertical or slanting joints require pasty types. A good result depends on preparation. Clean the joint and use a primer. Squeeze a length of plastic foam strips or polythene tubing into deep joints before filling them. Two-component joint fillers of this kind may even be tinted by adding PU color pastes. Their original color is gray or beige.

Bonding Difficult Material

Due to its simplicity, glueing is one of the favorite joining and mounting techniques. Sometimes, you may, however, get a rather unpleasant surprise, which is usually due to one of two causes:

The first reason may be that the two materials to be bonded together are not only slightly attacked by the solvent contained in the glue but completely dissolved, or the two materials will not stick together because the glue does not harden, which is the second reason for a failure. The reason for the latter is that both materials are only slightly porous or even nonporous and thus

Trouble caused by solvents

prevent the solvents contained in the glue from volatilizing, so that the glue remains in its liquid state, just as it does in its tube. In both cases, there is only one remedy: use a glue that is free of solvents. Besides epoxy resins, PU resins are gaining more and more importance in this field due to their excellent adhesion to nearly any material.

Later covering of a swimming pool with tiles

Thixotropic two-component PU compounds can, for instance, be used for bonding ceramic tiles or even glass to glass fiber. Fixing tiles to polyester swimming pools or shower stalls made from polyester is another example. The properly activated compound is dispersed on the base with a comb spatula. The characteristic increase in viscosity as gelling begins enables you to press the tiles into the thin coat of glue, where they will instantly stick without sliding down. You will very soon find out how long you have to wait to put the tiles in place and to work continuously. Such types of adhesive also prove very suitable for fixing polystyrene foam, which is attacked by most solvents, onto nonporous surfaces.

Even PVC panels can safely be fixed with PU glues, after a coating with one-component PU primer.

Molds and Casts

PU-based two-component casting compounds are also suitable for casting massive casts or molds, which may be either stiff or as elastic as rubber. Such compounds are frequently used as mold-making materials for plugs used for prefabricated concrete elements and for matrices for relief concrete by precasters, as this material is self-releasing against hardening concrete.

Modelers and do-it-yourself fans regard these compounds as a most versatile material for many purposes. Small or even bigger casts can be taken from the hard and tough compounds, which are normally used for recessed plugs. They are cast in molds made from sealed plaster, silicone rubber, wood, metal or polyester, which must be treated with special release agents for PU resins, recommended by the manufacturer of the resin. Handles with cast-in mounting bolts, model parts, sculptures and similar objects can be cast from this material.

Use special release agents

The elastic types (matrix compounds) are a less-expensive substitute for silicone rubber for the production of molds for undercut moldings and reliefs. Unlike silicone rubber, these PU compounds are not self-releasing, so that the use of release agents is indispensable.

Such a mold can be used for casting moldings from any kind of liquid plastic resin. They should be allowed to cure for about three days before they are used for the first time. Unlike silicone rubber, PU compounds do not reproduce extremely fine structures, such as fine wood grain, as release agents are required both when the mold is taken from the master model and when casts are taken from the mold. Very fine structures are partly obscured by the release agent.

Two-Component PU Lacquers

Due to their excellent adhesion, high elasticity and their resistance to abrasion, weathering, water and chemicals, two-component PU lacquers (which are also known as DD lacquers because of their two components, Desmodur and Desmophen, developed by the Bayer company) are highly suitable for painting highly stressed surfaces. They are used for painting boats, sealing inlaid floors and furniture and, as already mentioned, for the sealing of molds, where they score by the ease with which they can be sanded and polished.

Easy to sand and polish

Two-component PU lacquers are generally clear or only slightly yellowish, but the enamel industry also supplies opaque types. When buying lacquer, it is advisable to tell the salesman what you are going to use the lacquer for in order to get the most suitable type for the purpose.

Two-component PU lacquers contain solvents, which you can tell from the smell. The contents of solvents must also be considered when these lacquers are applied. In order not to be forced to accept too long hardening times, these lacquers are quite often allowed to prereact for about half an hour to one hour after the careful mixing of the basic component and the hardener.

Long pot life   In most cases, the pot life is fairly long and allows you to make quite large mixes for painting large areas, without getting into trouble due to lack of time. The pot life is also extended by low temperatures. Once applied in a thin coat, the two-component PU lacquer will harden fairly quickly and will normally have completely cured within four to six hours at 68° F (20° C). As a rule, such a lacquer will be hard enough to resist mechanical strain only after twenty-four hours, if the hardening is not accelerated by heat. After this time, the lacquer is hard enough to be impervious to fingerprints.

Heat must not be applied before the solvents contained in the lacquer have volatilized, which takes about half an hour to one hour, depending on the thickness of the film and on the type of lacquer used. Otherwise, the lacquer will gel on its surface too early, and the solvents contained in the still liquid material underneath will produce bubbles when they are trying to volatilize.

Heat-activated hardening is not only quicker but also improves the quality of the coat of lacquer. At a temperature of 176° F (80° C), which can be easily reached in the heating chamber of the paint shop of a garage, two-component PU lacquer will achieve its final hardness after only two hours. At 356° F (180° C), the lacquer will be quite hard within half an hour.

As a rule, an amateur will not be able to use heat for hardening unless he can use the facilities of a paint shop in a car body repair shop. But even if the lacquer only cures at normal temperatures, it will produce an extremely resistant film.

10.
Plastic Foams—
Especially
Versatile
Polyurethane Foams

Liquid PU foams have proved to be a most versatile material, which has been available for the do-it-yourselfer for some years and has recently become even more attractive because of simplified application.

Such foams are produced by mixing specified quantities of two-liquid components. Simultaneously with the interlacing of the resin, an expansion process takes place that turns the resin to a porous light foam. The classical chemical expansion process based on the chemical disintegration of some of the isocyanate with water, which leads to the splitting off of carbon dioxide (CO_2) and also sets free some heat, has already been mentioned. In addition, there is a physical foaming process caused by the evaporation of a liquid with a low boiling point added to the resin. This liquid already boils at normal temperatures and generates innumerable little bubbles filled with gas that expand the solid plastic material into a foam. For this purpose, special types of freon are used. This is a liquid that is quite well known as a pressurizing agent for spray cans and belongs to the group of fluorhydrocarbons. Both expansion processes can be used separately as well as in combination. The latter applies to most types of foam used in the do-it-yourself field.

Mixing Time and Pot Life

Short mixing time

The mechanism of the foaming process and its synchronization with the formation of the plastic structure by the interlacing of the two components is not only of academic interest but also explains two important conditions for successful use of the material: it has, unfortunately, is most cases, a very short mixing time, and this takes up some of the pot life which is only twenty to fifty seconds maximum; and it is necessary to work at a temperature of between 64° F to 72° F (18° C to 22° C) in order to make the foam expand to its full volume.

Maximum pot life fifty seconds

The short pot life is a specific feature of the system of chemical reactions involved and cannot be extended. As the foam is only pourable within this short pot life, special care must be taken to achieve rapid but thorough mixing. With normal types of foam, manual mixing is accordingly limited to mixes of up to 1 lb. (0.5 kg.). Bigger mixes require a frictional mixing attachment revolving at about 1,500 rpm and fitted in the chuck of an electric drill.

In cases of doubt, a mechanically mixed foam will always lead to better results. Its cellular structure and strength will be more evenly distributed.

Instant Shake Foams

These foams are an exception. Even though they are mixed by hand, their quality is as good as that of mechanically mixed foams. Instant shake foams are ideal for do-it-yourself jobs and their application is 100 percent trouble-free. The kit contains a plastic bottle that is only partly filled with the B-component, into which the A-component must be poured from a metal tin. A plastic nozzle is mounted on the plastic bottle, which is not completely filled so that the contents can be shaken. The nozzle must now be sealed with your thumb or forefinger, which is pro-

Trouble-free application

tected by thin polythene gloves included in the kit. The necessary thorough mixing of the components is achieved by vigorous shaking of the bottle for about ten seconds (approximately twenty shaking movements). Then, you can squeeze the ac-

This two-man sailing yawl, Brisia, was home-built from polyester resin and glass fiber mat in a rented female mold.

Garden furniture made from polyester resin and glass fiber—an attractive subject for the handyman with skill and the feel for design. First, a wooden master has to be carved, from which a female polyester and glass fiber mold is then taken; in this, the chair or table is molded. The edges of the laminate should be trimmed while the cast is still in the mold. The upper photograph shows a chair partly taken out of the mold—the photograph on the bottom of the page shows the finished chair made from glass fiber and a table of the same material. The addition of pigment to gelcoat resins allows you to make casts of any color you like. Painting is unnecessary.

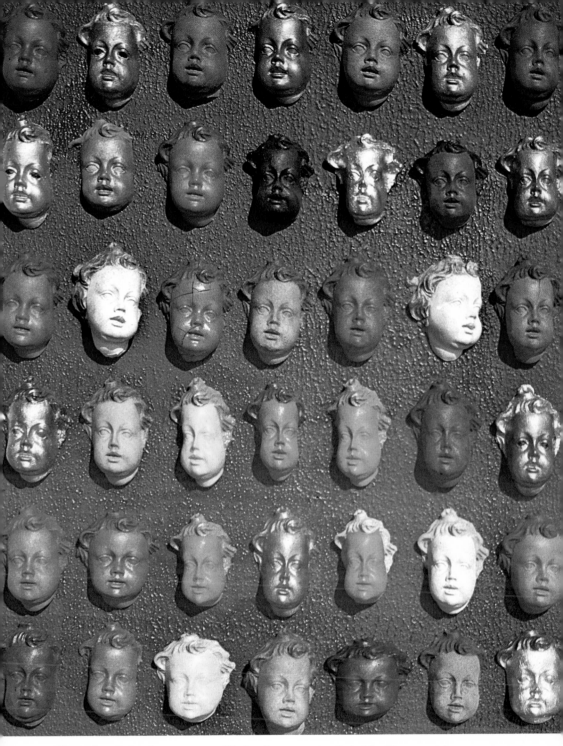

These cherub heads are cast in filled polyester resin: Suitable fillers are talc, fine quartz or sawdust and microspheres. Suitable mold-making materials are PU resin or silicone rubber (castings by K. B. Schoenenberger).

These "wood carvings" are polyester casts made in silicone rubber molds (casting by K. B. Schoenenberger).

Plastic furniture is getting progressively popular due to its interesting design features and because it requires little maintenance. Here you see two variations of this theme: foam chairs from resilient plastic foam with covers made from woven polythene strips. In the lower picture, you see a table and chair shells molded from integral polystyrene foam. Even the sideboards in the background are molded from the same material.

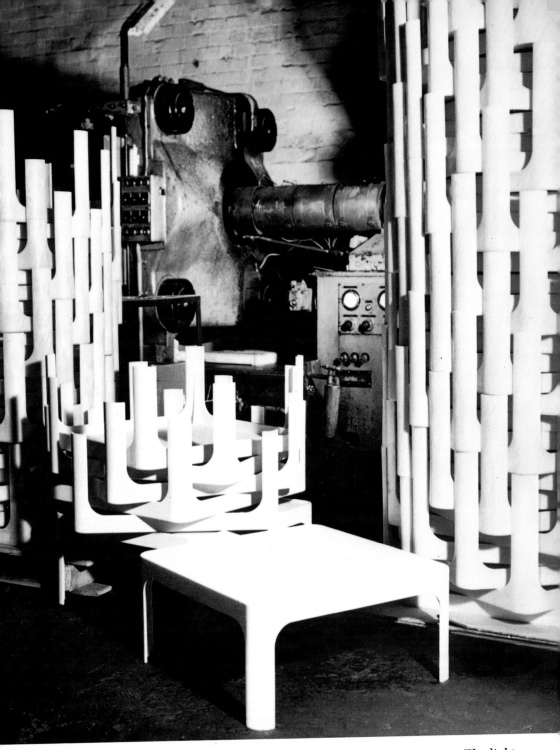

This is a machine molding table in integral polystyrene foam. The light porous core of the material is covered by a solid smooth "skin," lacquer finish will be applied later.

You can even build motor cars from plastics. This experimental car has a glass fiber chassis with a PU foam core. Racing cars often have bodies made from polyester and glass fiber.

Interesting effects can be obtained with strands of roving and polyester resin, as this lampshade shows. This sphere was wound on a machine but may be made by hand, given the necessary skill and patience. Take a balloon or a large plastic beach ball as a mold. You may wind the strand on the mold either dry or already impregnated with resin.

If you do not mind the work involved in making the molds, you can cast attractive chessmen from filled polyester resin (castings by L. Peinecke, Frankenthal, West Germany).

The technique of laminating allows infinite variety. This vase—not too difficult a job—impresses with its clear, slender shape. But the scope of embedding is no less extensive. Here, the parts of an old alarm clock were enclosed in clear polyester resin to make an attractive "conversation piece" or beautiful paperweight.

Impressive effects can be obtained by adding amber resin and crack-effect paste to polyester casting resin. The gelling resin cracks due to release of a gas from the crack-effect paste.

A great variety of materials (glass, metal, ceramics, chipboard) can be decorated with an enamellike glazing of tinted epoxy resin. The colors may be either clearly divided into sections or allowed to run into each other.

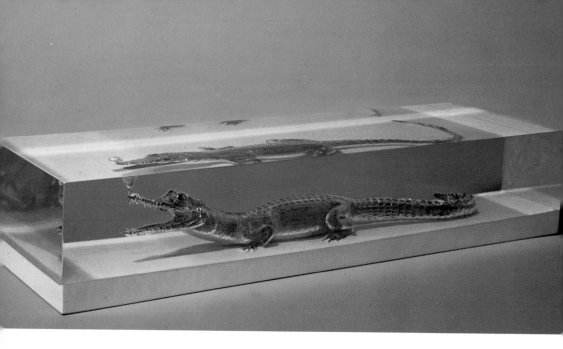

Embedded in clear polyester resin, many objects gain a new and special kind of attractiveness—in the right picture, you see two immortelles, which make an attractive object embedded in a high-gloss crystallike block of resin. When casting, you must carefully avoid air bubbles. Always keep in mind that large blocks may easily crack due to heat accumulation. You may prevent both risks by building up bigger castings, layer by layer and allowing each layer to harden properly before the next one is added. This technique was also used for embedding the crocodile in the picture above. The basic layer was tinted white to achieve a good contrast.

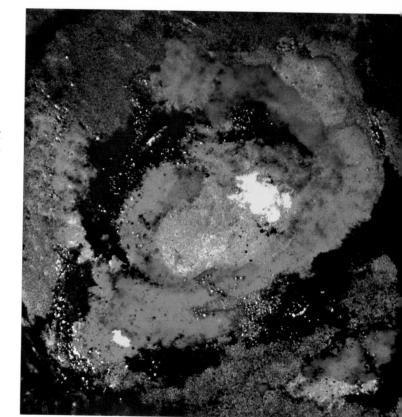

Oven fresh pictures would be a good headline for this impressive technique. The pictures are made by melting thermoplastics on a piece of sheet metal that was treated with release agent. Above: Composition of colored polythene granulate which was melted by heating and now gleams in bright opaque colors. Below: A molten picture from translucent polystyrene reminiscent of glass painting (by Max Hunziker, Zurich).

"Blue-Brown Sun" is the name the artist gave this relief. It was carved from a 24 in. x 24 in. (60 cm. x 60 cm.) polystyrene foam panel, the front side of which was protected with a coat of epoxy resin. Polyester resin poured from the reverse side dissolved the polystyrene foam core and filled the mold (by L. Peinecke, Frankenthal).

The German designer G. Neuhaus, from Frankfurt, created do-it-yourself furniture to be built from expanded polystyrene and high-gloss polystyrene panels—an attractive idea and one that is easy to accomplish.

Giving your home a Japanese touch. A partition wall built up from a wooden frame fitted with transparent-colored polystyrene panels looks rather exotic.

tivated mixture directly into the cavity by pressing the plastic bottle or allowing the foam to expand within the bottle and then escape into the cavity through the nozzle.

Important: The Temperature

The second important point is the correct working temperature, as the vaporization temperature of the expanding agent is reached too late in cold conditions. A large part of the heat generated by the chemical reaction between isocyanate and water provides the starting temperature for the evaporation of the freon and is lost if the two components are too cold. The two components may even suffer from a cold shock if the mixture has the right temperature but the mold is too cold. Here, again, too much heat is lost and the foam will not expand to its full volume.

The mold must not be too cold

An insufficient expansion may, however, be due to other causes. The foam may have been too old and have lost some of its extremely volatile propellant. On the other hand, very narrow cavities and molds with rough inner surfaces may also reduce the degree of expansion, because of wall friction.

But let us once again return to the question of temperature. We have learned that too low temperatures are deterimental. What about higher temperatures? They not only speed up the interlacing of the components of the resin but also activate the expansion process, as the volatilization temperature of the propellant has been reached before the chemical reaction of the components takes place. Therefore, you must be very careful when mixing the components, and you must be prepared for a reduction of the pot life. Instant shake foams react surprisingly fast at high temperatures. If you know about this, it will not lead you into any trouble. You should, however, consider that extremely high working temperatures may possibly lead to a coarser, and not so even, structure of the foam cells.

Which Foam for Which Purpose?

As with most liquid plastic resins, there is a great variety of different types of liquid PU foams. First, they are distinguished by their different weight per volume (specific gravity), which normally lies between 1.2 oz./cu. ft. to 4.0 oz./cu. ft. (12 kg./m.³ to 40 kg./m.³) as far as manual foams are concerned that are suitable for do-it-yourself purposes. With some special machine foams, which require expensive spraying equipment for

Specific gravity and cellular structure decide mechanical strength

DIFFERENT TYPES OF FOAM AND THEIR USES

SPECIFIC GRAVITY	FOAM STRUCTURE	FIELDS OF APPLICATION
12–15 kg./m.³* 1.2–1.5 oz./cu. ft.	Semirigid Open cells	Heat insulation in dry rooms, packing of fragile goods, stabilizing of thin-walled balsawood fuselages for model aircraft or model boats, noise insulation, landscapes for model railroads
25 kg./m.³* 2.5 oz./cu. ft.	Semirigid Closed cells	Filling of cavities and hollow parts of car bodies and chassis as a protection against corrosion and for strengthening and repairing rusted nonload-bearing hollow structures of car bodies, heat insulation at temperatures between −40° F to +248° F (−40° C to +120° C)
40 kg./m.³* 4.0 oz./cu. ft.	Rigid Closed cells	Buoyancy, making boats unsinkable, insulation against cold (repair of damaged insulations of refrigerators and deep-freeze equipment), stabilization of heavy duty model aircraft fuselages made from balsawood or glass fiber, landscapes for model railroads
40 kg./m.³ 4.0 oz./cu. ft.	Soft, resilient Open cells	Padding, cushions, upholstery, antidrumming of sheet steel structures
60 kg./m.³ 6.0 oz./cu. ft	Tenacious-resilient, high percentage of open cells	Shock absorbers, fenders for boats, fairly hard padding and cushioning
250 kg./m.³* 25 oz./cu. ft.	Very hard, rigid Open cells	Heavy-duty moldings

* Also available as instant foams.

two-component materials, the specific gravity lies between 6 oz./cu. ft. to 10 oz./cu. ft. (60 kg./m.³ to 100 kg./m.³) and may even reach 25 oz./cu. ft. (250 kg./m.³).

The structure of the foam cells is another important criterion. There are foams with open cells and others with closed cells, soft resilient types and semirigid and rigid types. Foams with closed cells absorb little or no water; those with open cells have a spongelike structure and, therefore, absorb water.

The mechanical strength of the foam partly depends on the specific gravity of the foam and, also, its cellular structure. The strength rises with the specific gravity and, to some extent, also with the hardness of the foam. Higher strength plus higher specific gravity can be achieved with standard foams by an overdosage of foam, i.e., by filling a cavity of about 5 cu. ft. (50 l.) with a quantity of mixture that would normally expand to a volume of 7.5 cu. ft. (75 l.) if its expansion was not restricted.

Increasing the density of foam is a much used technique and does more than increase its strength. It also guarantees a complete filling of the cavity, but you must pay for these two advantages with increased pressure in the cavity, which can reach fairly high figures of about 71 psi (5 atmospheres), so that the mold or cavity walls must be strong enough to withstand this pressure.

Color and Surface

PU foams can be tinted by adding 3 percent to 10 percent of special pigments (percentage based on the weight of the A-component) during mixing. With the milky or clear prepolymer foams, this will produce pure rich or pastel shades, while normal PU foams can only be tinted in dark colors, due to their own yellowish tints. Standard PU foams produce a thin glossy skin, which, however, is fairly susceptible to mechanical stress.

Special color pastes for tinting the foam

Safety Precautions

The safety precautions are the same as for solid PU resins. Look for extra good ventilation when you are working with white (prepolymer) foams. Wear goggles to avoid splashes of the hardening (B-) component from getting into your eyes. If it should still happen that some B-component gets into your

Splashes of hardener are dangerous

eyes, rinse the eyes with a 1.3 percent kitchen salt solution and see a doctor immediately. Splashes on your lips or drops of hardener that might have gotten into the mouth should immediately be washed away with dilute alcohol (for instance, beer). Do not swallow the liquid used for cleaning your mouth!

PRACTICAL ASPECTS OF PU FOAM

It is not possible to deal with all the interesting uses of PU foam within the compass of this book. Only a few that are of special interest are described in this chapter.

Packing

Fragile goods, such as cameras, mechanical or electronic precision instruments, etc., can be embedded in foam for safe transport.

The method is very simple: choose a cardboard box that is big enough to take the part to be packed and the protective foam. Put the fragile article into a polythene bag, which is tightly sealed with masking tape or a length of string. Pour some foam on the bottom of the cardboard box and allow it to turn hard before the polythene bag with the fragile part is placed on top. Make another mix and pour it on top to encapsulate the part completely in foam. A light packing foam is the best choice for this purpose, as it can be easily taken apart when the article is unwrapped. Alternatively, you can attach a length of string to the polythene bag long enough to serve as a ripping cord, which cuts right through the foam. The polythene bag can be easily removed as the foam does not stick to it.

Insulating Pipes

Noise is reduced, too

Insufficiently insulated pipes are a nuisance; they tend to freeze in winter and are noisy all the year round.

PU foam is a reliable and simple remedy. The pipe is encased in strips of chipboard, and the "shuttering" is then filled with a closed-cell foam. Later on, the chipboard is decorated with wallpaper or painted.

Expansion tanks of central heating systems freeze very easily. They can be encased in a large cardboard box, leaving enough space for an approximately 2-in. (5-cm.) thick layer of foam around the tank. The box is then filled with foam.

Making Boats Unsinkable

"The sea is not planked over," is a saying among landlubbers, and many a chastened weekend skipper has been made to see the truth of that. To prevent unscheduled dips, fill the cavities in your boat, which are of no use for anything else, with a closed-cell foam for buoyancy.

A 2.2-lb. (1-kg.) foam will expand to forty times its liquid volume if a 2.5 oz./cu. ft. (25 kg./m.³) type of foam is used and provides 17.7 lb. (39 kg.) of effective, useful buoyancy. A 4.0 oz./cu. ft. (40 kg./m.³) foam expands in a ratio of 1:25 and provides 50.7 lb. (24 kg.) of usable buoyancy per 2.2 lb. (1 kg.) put into the boat.

Buoyancy from foam

When calculating the required quantity of foam, the total weight of the boat, including equipment and the specific gravity of its structure, must be considered. If a glass fiber boat, for instance, weighs 353 lb. (160 kg.), about 132 lb. (60 kg.) of this weight must be balanced by foam, because the specific gravity of glass fiber laminate is approximately 1.6 g./cm.³ (.058 lb./cu. in. = 26.17 g./cu. in.) and only 220 lb. (100 kg.) of the total weight of the boat will be carried by the natural buoyancy of the building material.

In order to be on the safe side, the remaining 132 lb. (60 kg.) of buoyancy to be supplied by the foam should be increased by 30 percent to 40 percent, which, in this case, means another 40 lb. to 53 lb. (18 kg. to 24 kg.) of extra buoyancy. Add another 55 lb. (25 kg.) for a moderate rigging. This means a total additional buoyancy of 243 lb. (110 kg.), which means you require less than 6.6 lb. (3 kg.) of 2.5 oz./cu. ft. (25 kg./m.³) foam, or a little less than 11 lb. (5 kg.) of 4.0 oz./cu. ft. (40 kg./m.³) foam. The volume of about 3.9 cu. ft. (110 l.) of foam may be put into the air chambers within the hull, or the bottom of the boat can be filled with foam in order to provide extra buoyancy and additional strength.

Sandwich structure

The very good adhesion of the foam will tightly bond the original bottom of the hull and the extra floor and provide a very good shearing strength, so that the original bottom of the boat is transformed into a very strong sandwich structure, with heavy and dense outer panels firmly joined by a light core.

Car Body Repairs and Corrosion Proofing

Car body repairs and corrosion proofing are among the most interesting uses for instant shake foam. Hollow parts of the chassis and car body are subject to condensation, which leads to corrosion. The foam filling of such cavities is being adopted by the car industry. Several types of car on the market have foam-filled cavities to protect them against rust and reduce noise at the same time. Car body repair shops also use foam to reinforce beaten out dents for the prevention of rust and possible vibration.

Noise dampening as a free extra

• MORE SAFETY IN THE CAR. The most spectacular contribution to the discussion of "foam in the car" was made by the development department of the Adam Opel AG, the German subsidiary of General Motors. The designers found that box-type longitudinal members of the chassis will absorb impact shock much more evenly if they are filled with a fairly heavy PU foam. If such a box-type longitudinal member is exposed to a head-on crash, it will telescope like an accordion without tending to break or collapse, as hollow structures do (after which they are unable to go on absorbing the forces of the crash).

At about the same time, another German company, Voss-chemie (Uetersen near Hamburg), specializing in the technology of liquid plastic resins, was experimenting with PU foam as a means of preventing corrosion of hollow parts of car bodies and chassis. In the course of their experiments, the technicians of the company filled such cavities with a closed-cell PU foam in order to fight the rust by eliminating its causes: condensation and oxygen, which both attack the sheet metal from inside the cavities. In the course of these experiments, the technicians found that the foam not only eliminated rust in the cavities but, at the same time, effected a considerable increase in the bending strength of the hollow structures, without adding much to the weight of the car.

A "backbone" of foam for greater safety

The increase in structural strength was demonstrated by a test in which the door of a standard car—a typical example of a two-shell car body structure—was placed on two supports. A length of a steel girder was laid across the middle of the door to test the cross-breaking strength of the door by means of a giant weighing pan suspended from a steel girder. While the unreinforced standard door collapsed under a load of 1,069 lb. (485 kg.), another door of the same type, which had been filled with foam, only bent slightly. The bending slowly increased the

High cross-breaking strength

more the load was increased. It took 3,296 lb. (1,495 kg.) to make this door collapse.

This test was executed with a great number of doors of different types and makes of cars and always proved that the strength of doors was increased two or three times. Moreover, the foam-filled doors always showed a much more even absorption of the load.

Even if only the lower third of such a door was filled with foam, there was an increase in strength of up to 50 percent. Although complete filling of two-shell hollow parts increases the strength by 200 or even 300 percent, the doors of most cars can only be partly filled with foam due to the cranking mechanism for the windows.

Other tests with foam-stabilized cars proved that the damage caused by minor accidents—which generally costs a fortune to repair—is very much reduced, so that the foam may very soon pay for itself. Furthermore, it turned out that filling the cavities with foam reduces the noise within the car to a noticeable degree.

Plans for foam insulation are available for many cars, showing the position of the cavities to be filled and the sites of injection.

If you are afraid of drilling several dozen holes into your car for injecting the foam, relax. With most cars, the cavities can be injected merely by taking off the inner lining of the car body, so that no drilling is required at all. The VW "beetle," for instance, only needs a single ⅜-in. (10-mm.) diameter hole to be drilled. With other types of cars, it might be three to four holes. Where and how to drill these holes is clearly shown in the plan for each type.

Depending on the type of car, foam treatment costs from $60 to $145 (£25 to £60)—not too exorbitant for a process that provides reinforcement, protects your car against corrosion and perceptibly reduces the noise inside the car. This, at least, is the cost if you do the job yourself. If you have the work done by a professional, there will, of course, be labor costs. Even though the strength of the car is considerably increased, foam injection does not have much effect on the weight of the car. A medium-size car will only become 15.4 lb. (7 kg.) heavier by foam treatment.

• HOW TO INJECT THE FOAM. As mentioned above, it is sometimes necessary to drill some holes in order to be able to inject the foam into the cavities of the car body or chassis. Generally two ⅜-in.

(10-mm.) diameter holes are required, one of them for injecting the foam, the other one to allow the air contained in the cavity to escape.

A car-body specialist will be able to tell you where you can drill such holes without running into trouble. Alternatively, you can have a look at the plans that are offered by the manufacturer of the foam. A 2.5 oz./cu. ft. to 3 oz./cu. ft. (25 kg./m.^{3} to 30 kg./m.^{3}) instant shake foam with closed cells, which contains self-extinguishing properties against fire, is most suitable for this purpose.

Allow to dry before the foam is injected

For hollow parts under the doors, which are particularly subject to corrosion, the first hole should be drilled about 4 in. (10 cm.) behind the point where the hollow profile meets the vertical post to which the door hinges are attached. The second hole should be drilled about 4 in. (10 cm.) in front of the point where the hollow sill meets the opposite vertical post containing the lock. Remove the metal cover of the sill before drilling the holes into the sill. If you think that moisture might have settled in the cavity, allow the car to stand in the heated garage overnight with the doors wide open.

Drying can also be effected by heating the sill from the outside with a fan heater. Then, cover the interior of the car with a plastic sheet in order to protect it from foam splashes, which, by the way, can be removed only while tacky. Then, the two components of the foam are poured together into the plastic bottle and vigorously shaken. Once the mixing is finished, the injection nozzle is placed in one of the two holes to inject the foam. After two or three minutes, a small foam "mushroom" will rise from the second hole, which serves as an air vent, and show that the cavity is completely filled. In order to make sure that the foam really fills the farthest end, you may block this hole for a short time with your thumb, protected with a polythene glove. This leads to a slight compression of the foam and the resulting pressure will press the mixture into the remotest cavity. Then, remove the bottle with the nozzle and cut the foam flush at both holes, with a sharp knife. Give the foam a thin coat of polyester filler and screw the metal cover of the sill in place again. That's it!

Remove rust with compressed air

When filling longitudinal hollow parts on the chassis with foam, you should remove loose rust inside with a shot of compressed air from the compressor in the garage, after you have drilled the necessary two (not more!) holes. This blows the loose rust away, and the cavity is ready to be filled with foam.

For sealing the drilled holes, you can use plastic plugs, which

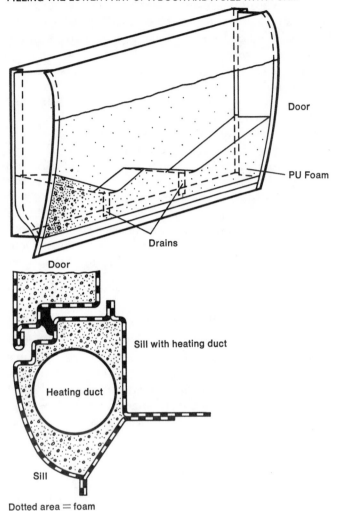

Door

PU Foam

Drains

Door

Sill with heating duct

Heating duct

Sill

Dotted area = foam

can be obtained from a car accessories shop, but you may also seal them with glass fiber filler or standard polyester filler. It is essential to ensure that the foam fills the hollow part completely for its full length. To this end, support the front of your car so that the profile is slightly raised towards the ventilation hole, because the foam has a natural tendency to rise.

Some parts that are easily attacked by corrosion can even be reached without drilling. Hollow double-walled parts of the car body in the back of the car, such as the rear wings, become accessible from inside after removal of the inner panels, which are only held in place by a few screws. The hollow parts above the rear fenders or mudguards can be reached by removing the covers for the rear lights. Here, the foam is injected through the hole for the electrical wiring of the lights.

Support rusted parts   If parts of the car are rusted through, instant shake foam is a great help too, unless load-bearing parts are partly eaten up by corrosion. Inject the required quantity of foam through the rust hole and press a piece of plywood or Formica against the thin sheet metal, which otherwise may not be strong enough to withstand the pressure of the expanding foam. In order to allow the air to escape, a ventilation hole of approximately 1/16 in. (1.5 mm.) must be drilled somewhere at the far end of the cavity (possibly drill from the inside). After the cavity has been filled, there may be areas of exposed foam, where metal has rusted away. This is now coated with glass fiber filler or laminated with a layer of polyester resin and glass fiber mat (refer to previous chapter). Apply filler paste, smooth, sand and spray with lacquer from an aerosol can—finished!

Noise Insulation

PU foams are also a most suitable material for noise insulation. They are ideal whenever parts with an intricate shape have to be insulated against the transmission of noise. In case of very loud noise, however, only one type of foam will help, which must have two special properties: elasticity and a fairly high weight.

Therefore, resilient foams with a specific gravity of approximately 7.5 oz./cu. ft. (75 kg./m.^{3}) are used, to which enough dry baryte is added to bring the specific gravity to about 10 oz./cu. ft. (100 kg./m.^{3}).

The fortified PU foam is then poured between the wall and shuttering, which is covered by a plastic sheet. This is self-releasing and is removed as soon as the foam has turned hard. Foam padding must always be fitted on the side of the wall where the noise originates. If all you want is to prevent sheet metal structures from drumming, 4 oz./cu. ft. (40 kg./m.^{3}) will do, but make sure that the foam is of the open-cell type.

Unlagged water pipes touching walls produce noise, too. A perceivable dampening of this noise can be achieved by encasing these pipes with chipboard and filling the cavity between the pipe and the case with a semirigid closed-cell-type PU foam of specific gravity of 2.5 oz./cu. ft. (25 kg./m.^{3}).

The same method may be used to insulate simple partition

walls of uprights and fiberboards. Filling the cavities of such structures provides good insulation against heat and noise.

Insulation Against Heat and Cold

At the present time, PU foam is the most powerful insulating material against heat and cold. This is demonstrated by modern refrigerators with their astonishingly thin walls. Having the same overall dimensions as refrigerators of the old type, which were insulated with other materials (stone wool, for instance), the modern refrigerators provide much more room inside.

This property is most valuable. It for instance enables us to build a portable garden bar with this foam. It consists of a rectangular box made from melamine-coated chipboard, into which bottles, treated with release agents, are placed. Use bottles of different sizes and fix them in the box so that there will be a 1-in. to 2-in. (2.5-cm. to 5-cm.) gap between the bottoms of the bottles and the floor of the box, as well as between the bottles. The bottles are best fixed in a chipboard lid, into which they are glued just at the point where their neck begins. This lid is then placed on the box after the quantity of foam required for filling it was poured in place in the liquid state.

How to insulate a garden bar with PU foam

The expanding foam tightly encapsulates the bottles and finally reaches the lid, which must have been treated with release agents or coated with plastic film or foil to prevent the foam from sticking to it. Do not forget to drill some ventilation holes into the lid in order to avoid an air cushion, which hinders the expansion of the foam.

Allow the foam to cure for a few hours before removing the bottles. Then, cut the actual cover of the bar box from melamine-coated chipboard and use the provisional cover, which was used for fixing the bottles, as a sort of template for where the full-size holes for the bottles have to be cut. This lid is fixed on the box with wooden dowels or screws or is just glued in place. You may either leave the necks of the bottles free or, if it is intended to keep the drinks cool as long as possible, make special covers for the necks, too. This can be done in two ways:

a) Laminate glass fiber caps, for which the lower half of the bottle is used as a mold. Remove these caps from their molds and pour a small quantity of liquid foam mix into the cap. Hold the matching bottle upside down into this beaker and try to fix it exactly in the center of the beaker. The foam will then evenly encapsulate the neck of the bottle, which must have been pretreated with release agent. Allow to harden and remove the bottle. Cut any protruding foam away, and trim the beaker to the desired length. Apply filler to achieve a smooth finish, sand and apply enamel from a spray can. These covers should tightly fit the holes in the lid of the box, which must be cut slightly bigger. This technique has a great advantage: each bottle can be easily reached by only removing its own cover.

b) Make a second box from melamine-coated chipboard. The height of this box must correspond to the length of the protruding necks of the bottles. Put the bottles into the lower box, and carefully treat both the bottles and the chipboard cover of the lower box with release agent. Put the upper box on the ground, with its opening facing upward, and pour the foam mix into it. Then, turn the original lower box with the bottles fixed in their holes and put it upside down onto the box with the rising foam. If you have put the correct quantity of foam into the box, the necks of the bottles will be tightly embedded in foam, so that the detachable cover for your bar box provides perfect insulation.

Allow the air to escape Ventilation holes are essential to allow the air to escape, as the foam expands. This is the reason why you can also work the other way round and place the flat box-type lid on the base with the bottles to be pretreated with release agents. In this case, the cut edges of the lid should be prefitted with melamine strips, which are glued in place and treated with release agent.

Any slit between the upper and the lower parts of the bar box should be sealed with masking tape before the foam mix is poured or injected through holes drilled into the top of the upper box. They can later be used for mounting a handle with bolts and spreading brass dowels or they can be hidden under a decorative sheet metal or plastic cover glued in place.

Filling Holes in Brickwork

Wherever pipes, cables or other kinds of electrical wiring have to be led through existing walls, cold and moisture will easily creep into the building. Quite often, such holes are left by electrical fitters or plumbers who do not consider filling them in as part of their job. Such a hole can be easily sealed with a bottle of instant shake foam of the correct size if you use cardboard templates as a temporary sealing on both sides to prevent the foam from runing away. Once the foam has turned hard, all you have to do is scrape away a thin layer of foam and replace it with a thin coat of cellulose filler.

Modeling

Model makers can solve many of their problems with liquid PU foam. For instance, it is quite easy to cast the complete scenery for a model railway, including a very long tunnel (made by fixing a length of a semicircular corrugated tubing on the base). The contours of the hills and mountains are shaped by a sort of tent made from polythene film or foil stretched over dowels; the latter are fixed in holes drilled into the baseboard. The outer edges of the film are fixed with thumbtacks or drawing pins and sealed with masking tape.

Tunnels and mountains from liquid foam

A hole is cut near the top of one mountain and the liquid foam mix is then injected through this hole. It expands under the tent and takes the shape of it. The final contours can now be molded with one's hands through the plastic film or foil, which, therefore, must not be stretched too tightly. For this purpose, the foam can be tinted in a suitable basic color (green or brown). Once the foam scenery has turned hard, the film is peeled off, and the scenery is now modeled in its final shape by sanding or cutting the foam with a bread knife with a serrated blade, a saw blade, a whip saw or a piercing saw. The project is finished by painting or glueing colored sawdust into the foam to represent grass. Generally, a 2.5 oz./cu. ft. (25 kg./m.³) rigid foam will be sufficient for this purpose, but quite often even a 1.2 oz./cu ft. (12 kg./m.³) type of foam will do. Additional tunnels can be made into the foam with one's hands or with a piercing saw. There probably is no other technique that has a wider appeal. Model boat fans, as well as aeromodelers, may use liquid foam for strengthening lightweight, thin glass fiber structures. Here, it proves to be most advantageous that the foam sticks to glass fiber surfaces.

Too much foam may burst fuselages or hulls

It is most important not to put too much liquid foam into hollow fuselages or hulls. They may burst.

It nearly is a life insurance for your radio control equipment if you wrap the delicate electronic parts in polythene film or foil bags and then embed them in the fuselage with a resilient type of foam, which, at the same time, absorbs the mechanical oscillations of the engine and also protects the radio control equipment against them.

Wings for heavy duty model aircraft, such as control line

combat models, can be easily made from a 1.2 oz./cu. ft. (12 kg./m.³) semirigid foam in a sort of serial production. The skin of the wings is made from 1/32-in. to 3/64-in. (1-mm.) thick balsawood or veneer wood of half the thickness, which must be fixed in a sheet aluminum jig bent to the correct shape. Glue a thick end rib into the open ends of the wing, and cut the biggest possible hole into the upper rib for pouring the foam mix into the wing. Do not forget a second hole, which can be smaller, for escaping air.

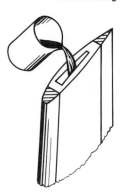

PU foam for aeromodeling

Filling a hollow wing with liquid foam

Pour the liquid foam mix into the wing, which is held in a vertical position. Once the whole quantity is poured into the wing, the latter should be held at an angle of only 45° which is reduced more and more the more the wing is being filled by the expanding foam, which finally expands through the filling and air escape holes. Cut the protruding foam away and allow the wing to rest in its jig for an hour. The foam core and the skin of the wing now form an integral part. The foam sticks perfectly to the wooden skin and has even got a slightly denser structure near the wooden skin, while its core has remained light and more porous, so that this structure can be called a perfect sandwich cast in a single process.

Extremely strong wings can be achieved if the thin shell of such a sandwich wing structure is fitted with a leading and trailing edge from soft balsawood, which is glued to the skin before the hollow wing structure is placed in its supporting jig.

Model boats become unsinkable

Model boat builders make their boats unsinkable by filling the cavities within the hull with a light closed-cell-type foam in order to prevent the loss of their expensive radio control

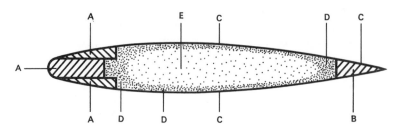

CROSS SECTION OF A SANDWICH-WING

The letters refer to the components of a foam-filled model wing.
A = built up leading edge, B = trailing edge, C = balsawood shell,
D = dense outer zone, E = foam core with normal weight and bigger pores.

equipment and of the boat, which may also be valuable. Even for filling the floats of model aircraft, the foam has proved an important safety factor, as it even keeps damaged floats afloat.

Reinforced Surface

While polystyrene foam is easily attacked by the styrene contained in polyester resin and, therefore, requires a protective coat of white glue if epoxy resin is used instead of polyester, PU foam can be directly coated with polyester and glass fiber, providing the foam is not too fresh. Nevertheless, it is advisable to give the surface of the foam a thin coat of a quick-acting polyester resin mixed with some extra accelerator and hardener. The mix is brushed or rolled on and should have a pot life of approximately fifteen to twenty minutes. You may also apply a thin coat of polyester filler, which also cures very fast. Even glass fiber filler is suitable for reinforcing the surface of PU foam.

Giving the foam a coat of glass fiber (polyester and glass fiber) is a useful technique in many fields of repairs, such as car body repairs, but also when foam is used in a boat for extra buoyancy. Glass fiber provides a protective coating wherever the bare foam is exposed and may be affected by mechanical loads or strain. Above all, fiberglassing the buoyancy foam in a boat is an extra safety factor, as it provides a firm joint between the hull and the foam, which does not disintegrate even under high stress, so that the foam cannot come off under emergency conditions.

Casting an Integral Sandwich

A light, porous foam core with a strong reinforced surface does not, however, depend on subsequent fiberglassing or on filling a gap between two hard panels supported by a jig when the foam is applied. So-called integral or structural PU foams provide a much more elegant solution to this problem. Unlike thermoplastic types of foam with similar characteristics, integral PU foams can be used for do-it-yourself purposes, as they do not require special machines.

Integral or structural foams with a hard skin

Expanding in a female mold, these foams produce a solid, nonporous skin, which is about 1/32-in. to 3/32-in. (1-mm. to 2.5-mm.) thick, while the core of the molding remains light and porous.

This effect results from an intentional cooling of the reactive mixture as it comes into contact with the mold surface. The skin will have its greatest thickness if the temperature of the mold surface is about 50° F (10° C). At normal room temperature (68° F [20° C]), the skin will still be 1/32-in. to 1/16-in. (1-mm. to 1.5-mm.) thick and at 122° F (50° C), the skin is so thin that no actual strengthening of the surface is achieved. Beside the temperature of the mold, its wall thickness has some effect on the thickness of the skin. With thin walls, between ¼ in. to ½ in. (6 mm. to 12 mm.) the skin will be thicker; a thickness of less than ¼ in. (6 mm.) will produce near-solid walls with very fine pores. The guidelines refer to thicknesses of 1 in. (25 mm.) and over.

It is quite an interesting fact that the thickness of the dense skin can also be controlled by means of the quantity of resin. If you put more resin into the mold than is required just to fill it, the surplus material will produce a higher specific gravity of the outer skin, while the core does not differ very much from material allowed to expand freely without compression.

Mixing time maximum twenty seconds

Structural foams allow the hobbyists a maximum of twenty seconds for mixing and remain pourable for about thirty seconds (pot life). The expansion of the foam requires about two to three minutes. After fifteen to twenty minutes, the casts can be taken out of the molds. Such foams are generally white, but tend to yellow when exposed to weathering in the open air. They may, however, be tinted, so that the yellowing will not be so striking. Between 10 percent and 20 percent of pigment powder (based on the weight of the A-component) can be added. The pigments are available in different colors, including wood shades. They must be completely dry.

In industry, such foams are used for padded instrument panels for cars (tenacious-resilient types), which have a leather-like finish when they come out of the mold so that they need not be covered with leather or any other material, door handles and arm rests for motor cars and even furniture (hard furnishings). But there also are various do-it-yourself uses, such as model-making, replicas of works of art, panels and toys.

Suitable do-it-yourself mold-making materials include wood,

glass fiber and silicone rubber for undercut masters. When casting molds with deeply structured surfaces, you must take care that the still liquid resin is evenly dispersed over the whole surface of the mold in order to avoid air bubbles and other faults. Oblong molds are best supported at one end, so that they are inclined at an angle of 10° to 15°. They are filled at their lowest point. But do not forget to provide a ventilation hole at the top end.

As the dense nonporous surfaces take up much resin, you will have to use more than the volume the mold would require at normal expansion rates. Generally, the quantity of resin required for a good cast with a "thick skin" should be two to five times the quantity required just to fill the mold. Allowance must be made for the extra pressure when building the mold.

Larger quantity of resin

Unlike conventional PU foams, integral foams are mixed by stirring the relatively flowable A-component into the much more viscous B-component. Additives, such as pigments and fillers, e.g., arenaceous quartz, are always premixed with the A-component, which is then mixed with the B-component by means of an electric drill fitted with a frictional mixing attachment (approximately 1,500 rpm). It is not advisable to mix such resins by hand. It is also essential to ensure that the mixing proportions do not deviate by more than plus or minus 1 percent.

The A-component is stirred into the B-component

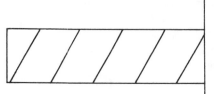

11.
Self-Releasing Flexible Molds from Silicone Rubber

As semi-inorganic polymers, silicones are of special interest to the chemist, but do-it-yourself enthusiasts and model makers appreciate the special properties of this versatile material. In addition to waxy release agents containing silicones—which, however, often involve adhesion problems in the subsequent painting of casts—there is silicone rubber. This is a most interesting material and is available in many varieties.

The main use of silicone rubber is in the production of self-releasing flexible molds for casts from liquid plastic resins. Owing to their excellent heat resistance, some types can even be used for the casting of metals with low melting points. A further use is the filling of expansion joints and sanitary joints. Silicone rubber has outstanding electrical properties and is resistant to weathering, heat and abrasion. This is why it is also used for embedding electrical and electronic circuits, thus providing a

Excellent electrical properties, resistance to weathering, heat and abrasion

perfect insulation, as well as protection, against moisture and vibration.

Flexible silicone rubber tubing and homemade socket couplings can be used to conduct hot gases or liquids if a flexible joint is required between two otherwise rigid pipes (for instance, with exhaust pipes for model engines). Molding compounds and highly resistant casting and embedding resins based on silicone are generally supplied as two-component systems, while jointing materials are one-component systems, which are cured by the moisture in the air.

Two-component materials are mainly gray, beige or whitish, but may also be brown to blackish or colored. In any case, they are fairly viscous.

The amount of hardener that has to be added allows a fairly wide variation and, thus, influences the pot life and the time required for vulcanization (curing time). That means that the same material mixed with 1 percent of hardener may have a pot life of forty minutes at 68° F (20° C) and a curing time of about four hours. By doubling the percentage of hardener, the pot life shrinks to twenty-five minutes, while the curing is completed within two hours. Another 1 percent of hardener leads to a pot life of fifteen minutes and a vulcanization within one hour. A total amount of 5 percent of hardener (based on weight proportions) leaves the material usable for only eight minutes and makes the resin cure within fifteen minutes. The mold will last much longer if it is kept at room temperature for four to six days before being used for casting.

Variable percentage of hardener

As already emphasized, silicone rubber is self-releasing when used with plastic resins, while being perfectly self-bonding. This enables you to cast a mold in several layers and allow it to be cured layer by layer. Where greatly undercut masters require a two-piece mold, the halves can be prevented from sticking together by treating the contact surfaces with a thin coat of soapy water or two-component PU lacquer. A thin oil film will have the same effect.

In order to achieve a perfect bubble-free inner surface, a thin gelcoat, consisting of silicone rubber that has been very well mixed with hardener, should always be brushed on the surface of the master. Then, make a fresh mix and pour it onto the gel-

Make the mold at least ⅜-in. (10-mm.) thick

coat. Make sure your mold is at least ⅜-in. (10-mm.) thick all around to prevent its being torn when the cast is removed.

For extra strength, embed a layer of woven roving (glass fiber) in the last layer of silicone rubber, which forms the outside of the mold. As mentioned earlier, cured silicone rubber molds can also be embedded in plaster. This reduces the consumption of silicone resin, which is fairly expensive. When working with polyester resin, the self-releasing properties of the mold can be improved and preserved by taking two casts in natural or paraffin wax before the mold is filled with resin for the first time. The use of wax leads to a very thin additional film, which aids release, protects the mold surface and improves its resistance to the styrene contained in polyester resin. This increases the accuracy of the castings by reducing the danger of swelling of the mold.

Preheating the mold to 140° F to 176° F (60° to 80° C) has also proved helpful when taking polyester casts in silicone rubber molds. The higher temperature leads to faster gelling of the resin that is in contact with the mold and leads to a 100 percent nontacky surface of the cast, as the styrene can no longer penetrate into the silicone rubber. (Special precautions have to be taken when casting PU foam in a silicone rubber mold. PU foams stick to nearly any kind of material and, therefore, exhaust the self-releasing properties. To avoid any trouble, special release agents for PU foams are indispensable after a few castings.)

Metals with low melting points, e.g., tin (450° F [232° C]), or lead (62° F [327° C]) or alloys of both, can be cast in special types of silicone rubber with extra high resistance to heat and subsequent intensive cooling as soon as possible after the casting is completed. Casting temperatures above 572° F (300° C) are very critical. This is why pure lead should not be used for casting if more than a single casting is required.

The great advantages of silicone rubber as a molding compound are that (a) the molds can be used over and over again and (b) undercut casts can be taken out of the mold without damage to either. As mentioned previously, the mold should always be allowed to cure for a couple of days and be tempered at moderate heat.

A very smooth surface of metal castings can be achieved by dusting the inner surfaces of the mold with ultrafine metal powder or silicone carbide dust. This trick also helps to reduce the impact of the heat on the surface of the mold.

Another most useful application of silicone rubber is the securing of nuts and bolts that are exposed to vibration and might otherwise get loose. Putting some silicone rubber on the thread, when the nut or bolt is screwed in place, prevents spontaneous loosening, but does not prevent them from being unscrewed if necessary. **Securing nuts and bolts**

The good heat resistance of silicone rubber makes this method also work with parts exposed to heat (e.g., engines), unless the silicone rubber is distorted by excessive heat (i.e., more than 572° F [300° C] maximum). Finally, there is still one application left that is worth mentioning in connection with do-it-yourself jobs: the sealing of sanitary joints, for which one can even buy a special formula with bactericidal additives (Bostik), and the filling of expansion joints. The application of silicone rubber in these cases does not differ from working with similar grouting materials, based on thiokol or polyurethane, which were dealt with earlier (see pages 137–139). Smoothing the injected paste is very easy: use a wet spatula or your wetted fingertip. **Sealing joints with silicone rubber**

It should be mentioned that there are considerable differences in the prices of one-component grouting materials, which reflect different degrees of resistance and elasticity. Therefore, it is wise to study the specifications and instructions on the container or the technical data sheet before buying a product required for a special problem. Extremely cheap silicones are probably based on other cheaper resins and contain only a small percentage of silicone rubber.

If compared with thiokol or PU pastes, one-component silicone rubber cures more quickly and is often more resistant to changes. Good silicone materials withstand a temperature of 302° F (150° C) over a long time without any trouble and can even resist higher temperatures for a short period. They also remain elastic at low temperatures of about −58° F (−50° C) and stay soft in their noninterlaced state, so that they can be applied at low temperatures without any problem. **Good materials withstand temperatures up to 302° F (150° C)**

12.
Finished Products
and
Semimanufactured
Articles
from Duroplasts

Facings, laminated panels, printed circuits

Duroplasts enter the do-it-yourself world not only as liquid resins but also, in their final hardened states, as finished or semimanufactured products. Well-known and widely used examples include corrugated glass fiber panels, serving as facings for railings on balconies or as a roofing material, and laminated panels, such as Formica, Micarta, Resopal or Ultrapas, which are used for tabletops and other furniture. Further examples are copper-coated resin-impregnated paper, which is a close relative of laminated decorative panels, such as Formica, Micarta, or plain laminated paper, which is used for mounting electric and electronic circuits. Last, but not least, there are compound materials, such as resin-impregnated wood fibers, that are molded to the desired shape by heat and pressure.

The latter consist of wood chips cut in a special way to preserve their fibrous structure. These chips are mixed with urea formaldehyde resins and pressed into moldings and profiles. In this process, the surface of these parts is coated with a melamine-impregnated decorative film or foil of the same kind as that used for the top of decorative laminates, such as Formica or Micarta. The material is available in two different types: the first one is intended only for indoor use; the second one is weatherproof. The latter can be easily identified by its reddish chipwood core, which is exposed when the profiles are cut to the required length.

Even though the binding resin is fairly weather resistant and even though the melamine surfaces are completely weatherproof, the wood fibers not only retain their fibroid structure, which adds stiffness to the cast, but also retain hygroscopic tendencies. This is why all cut edges must be sealed with a two-component epoxy filler, which is available in several colors matching those of the decorative panels. The filler seals the open edges and prevents moisture from penetrating into the wood fibers, so that the casts cannot warp. Resin-impregnated wood chip profiles are used as facings for balcony railings, as wall panels and window sills both inside and outdoors, as skirting boards and handrails on staircases, as blinds for indirect lighting and as facings of gables and similar embellishments around the house. Such profiles are available in many attractive decors, and their surface resists smoldering cigarettes, all household chemicals and, to a large extent, even high mechanical strain, so that this material can be widely used for many do-it-yourself jobs, ranging from the facing of radiators for the central heating system or the facing of balcony rails and garage doors to the building of garden furniture or giving the bathroom a "wooden touch" with a water-repellent lining. The panels can be easily cut to shape and drilled with normal woodworking tools.

Chipboards bonded with urea formaldehyde resin are also related to molded chipwood. You can buy them with and without decorative films or foils on both surfaces. Their production on a large scale revolutionized the production of all kinds of furniture, as these panels, which were supplied in standard sizes, allowed a high degree of automated production. Contrary to molded chipwood, the chips for chipboards are cut at random, so that no special structure of the fibers is maintained. A high degree of compression, the resin binder and the fairly hard decorative covering layers are the reasons why the saw gets blunt quite soon if you are cutting a lot of chipboards or molded chipwood. If you plan a bigger job with such materials, you should always have a sharp spare saw in store or, even better, only use a more resistant hard-metal saw instead.

You will find detailed instructions for drilling and sawing

How to work with chipboard

hardened duroplasts in the chapter "How to Work Plastics—General Rules and Special Methods" on pages 230–261.

Some hints on mounting duroplast panels: always drill screw holes slightly too large. As such materials have a poor thermal conductivity, they may easily heat to fairly high temperatures and expand to a greater extent than the material underneath to which they are fixed. Oversize mounting holes allow a certain degree of compensation and prevent warping. Correspondingly large washers under the head of the screw hide the hole and, at the same time, ensure a more even load dispersion.

Laminated panels with genuine copper top

It is also worthwhile to make a note with regard to the decorative laminated panels. They are available in a wide variety of standard decors and finishes (imitation wood grain, a great variety of colors, tile-look, etc.). Also available is a special type of laminated panel coated with a layer of genuine copper, which makes for attractive effects but requires special handling.

Just like normal copper sheeting, these panels tend to oxidize on the surface, which turns black or produces the well-known patina. This can be avoided by giving the copper surface a protective coat of lacquer. You may either buy the panels already sealed or apply the lacquer yourself. In the latter case, it is important that you sand the metal surface with fine steel wool and do not touch it any more with your hands before the sealing coat of Zapon varnish (cellulose "dope") or two-component PU lacquer is applied, as the slightest trace of sweat can cause discoloration, which only becomes visible after the surfaces have been sealed. Grease from your hands can be detrimental to the adherence of the lacquer to the metal surface or produce fingerprints. Also, be careful to avoid adhesive stains that may not be removable or may distort the copper film (urea formaldehyde glues).

You may, of course, allow the copper surface to oxidate, which looks quite nice and also protects the copper. A third way of surface treatment is the etching of old engravings and ornaments into the copper surface, which is then sealed with clear lacquer.

The etching of the contours is done after all metal surfaces that are intended to stay have been sealed with an acid-resistant type of lacquer. As a rule, this cannot be done as a do-it-yourself job unless you have special skill and knowledge of printed circuits. But even if you think that you can do the job, it is still worthwhile to consider whether it would not be better to have the etching done by a specialist, as the material is quite expensive.

Boat building: After the hardening of the gelcoat, a glass fiber cloth or mat is applied with polyester resin. The actual skin of the boat shell is built up from a glass mat and resin laminate, which is here being rolled with a metal-disc roller.

Four different types of glass fiber for reinforcing polyester resin: fine linen-type fabric (top left) and light glass-fiber twill (top right), at the bottom from left to right: woven roving and chopped strand mat.

Garden pools, ranging from a small ornamental pond to a full-size swimming pool, can be laminated from polyester resin and glass fiber without any trouble. Such pools are durable and require virtually no maintenance. The plaster-jute method is recommended for lily ponds and larger pools with slanting sides. Jute fabric soaked with a thin mixture of plaster and water creates a mold on the roughly shaped soil and provides a solid membrane, which prevents the resin from penetrating into the ground. It is hard enough to allow proper rolling of the laminate in order to avoid air bubbles. The two lower photographs show the building of a swimming pool with a concrete base and sidewalls built from limestone bricks. The interior of the pool is then plastered and coated with three layers of polyester resin and glass mat. In the left picture, you can see the supporting pillars that reinforce the sidewalls against the pressure of both water and earth. On the right, the glass fiber lining, which is molded over the crown of the pool walls, is treated with a metal roller to remove air bubbles.

For casting deeply undercut sculptures, latex is an ideal mold material. The mold is built up by frequently brushing the master model with latex, which can later be stripped off from the master model, as well as from the following castings, just like a stocking. For casting, the elastic skin of the mold is supported by a two-piece plaster mold.

An exciting moment (see photo below): the home-built boat is taken out of its female mold, built up from fiber boards mounted in a framework of wooden formers. This kind of mold, however, only allows sharp-edged designs.

1. Here, corrosion has done a perfect job! First, remove any loose rust and unevenness!

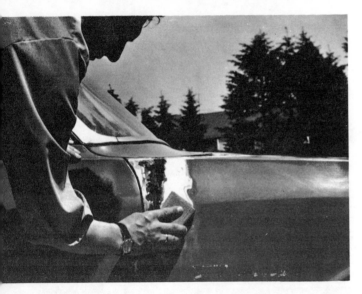

2. The best thing to do is to sand the metal around the hole down to the bare metal and then give it a thin coat of polyester filler, which is slightly sanded after hardening.

3. Tear a patch of mat to the required size and mix the required quantity of resin with hardener according to the proportions given by the manufacturer of the resin.

4. *Evenly impregnate the strip of mat on a solid support covered with plastic film or foil. Then, transfer the mat on the film to the area to be mended.*

5. *Peel the film or foil off carefully and with even tension. Dab the mat with a brush to remove bubbles.*

6. *Allow the repair to harden. Then, sand the surface, apply filler and sand again. Finally, apply a coat of lacquer and the car looks perfect again. Such a repair generally lasts longer than the car body.*

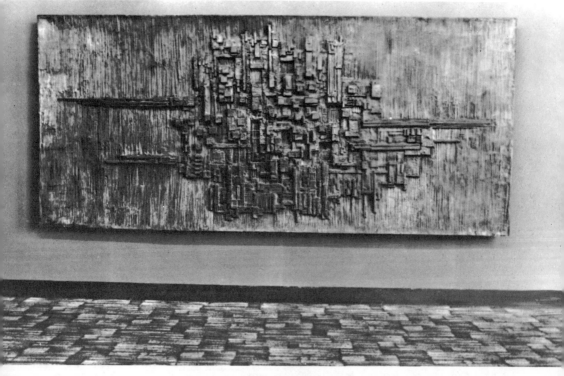

Plastics in the home, industry and the arts: Room divider built up fro
enameled integral polystyrene foam, sports car body (glass fiber) and larg
wall relief (polyester resin, glass mat and polyester filler).

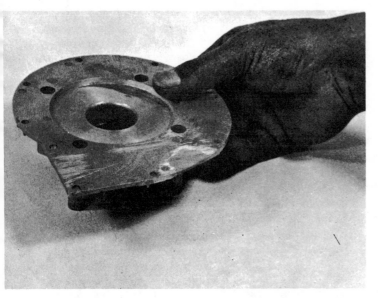

Polyester or epoxy resins filled with metal powder are ideal for repairing damaged metal parts. In the photograph below, a mixture of resin and metal powder, which also contains hardening powder, is cast onto the masked part. After hardening, the surface is filed flush, and the part is serviceable again.

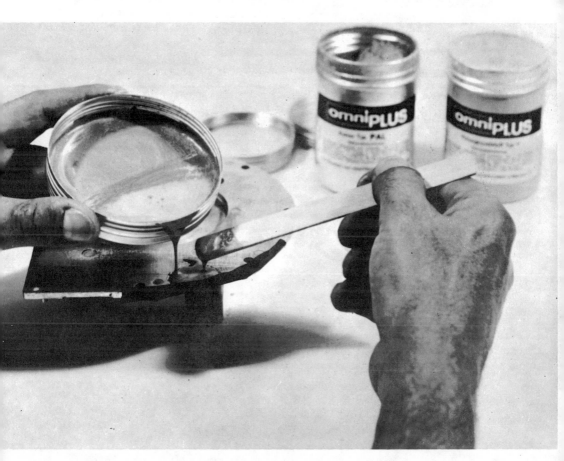

To the naked eye, microspheres look like fine dry sand. This microphotograph reveals their real structure: They are microscopic hollow spheres of glass, ceramics or phenolic resin. They are an ideal filler for polyester resin and, combined with wood meal, make perfect imitations of natural wood. Proper mixing of the fillers with the resin is a prerequisite condition. Before adding the hardener to the resin, you should allow the mixture to stand for several hours in a covered container to make sure that the wood flour is well saturated with resin and that there will be no bubbles later on. Below: Cast in imitation wood. A mix from wood flour, resin and microspheres is cast into a silicone rubber mold.

PU integral foam is an ideal material for making copies of reliefs. Even though silicone rubber is self-releasing, it is highly recommended to use release agents when making PU casts, as the self-releasing properties will soon deteriorate. In this picture, you see a casting which stuck to the edge of the mold that reveals the porous interior structure of the integral foam. Below: Elastic polyurethane. A mold that was taken from a wooden panel and a copy of this panel—both cast from Flexovoss K9.

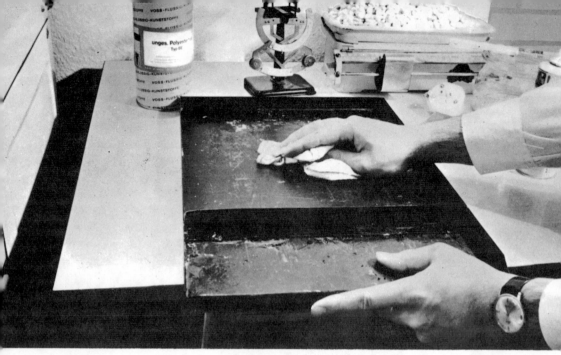

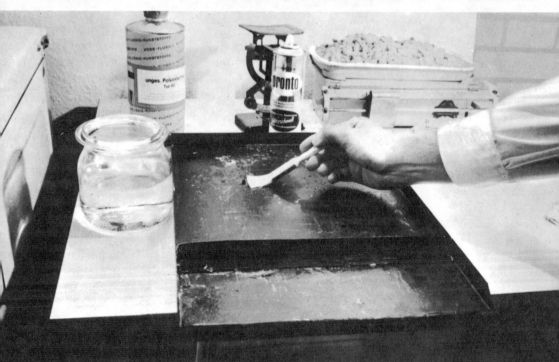

Artificial stone tiles can be made from small pebbles and polyester resin: A flat-tin box treated with release agent before a thin gelcoat of activated resin is brushed in and built up.

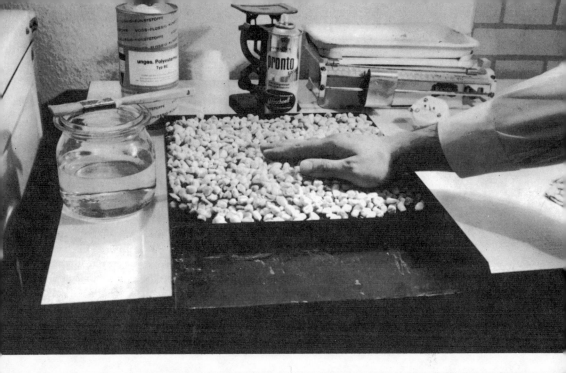

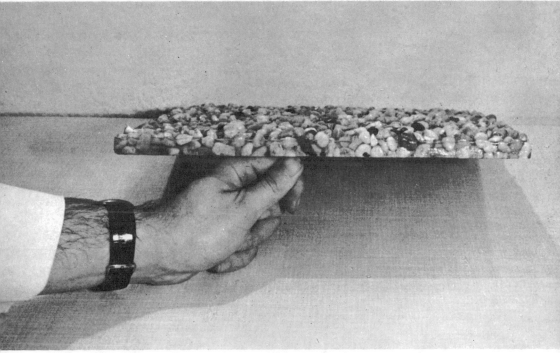

The pebbles are dispersed on the hardened gelcoat and then bound together by pouring on activated resin. Below: The finished tile removed from the mold.

Reinforcing hollow parts with instant shake foam is a practical way of repairing rust-damaged car bodies (left). The process also increases the safety of each car. The injected foam makes the weak sides of the car more resistant. This reduces the risk of damage when a collision occurs. The two cars will tend to slide off each other. Above all, the injected foam protects the car against dangerous corrosion inside the cavities and, also, reduces the noise. The picture below shows which areas can be stiffened in a Volkswagen "beetle."

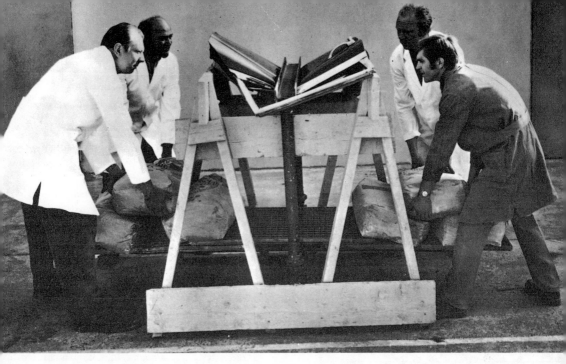

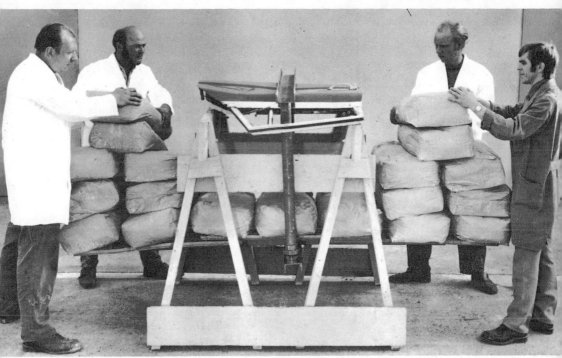

This buckling test demonstrates the increase in rigidity by filling hollow structures with foam. Resistance to buckling is more than three times greater in foam-filled doors.

Loose gravel in front of a garage quite often causes trouble, which can easily be overcome with liquid plastic resin. The gravel is mixed with one-component PU resin in a mixing machine, evenly dispersed on the surface and then smoothed with a float. Once the resin is cured, the pebbles are safely bonded together, and rainwater can still run off.

Plastic veneers are a most practical material. They fold easily around sharp edges, as you can see in this picture.

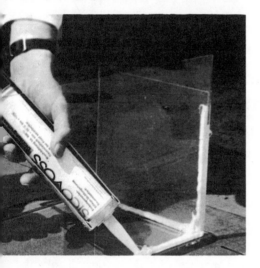

Glueing pads, like the practical Bostik Pads, provide very strong joints and can also cover a slightly uneven surface when mounting wall panels. In the left picture, the glue is applied, while the right photograph shows how the foam strip is pressed down. Below: Thermoplastic materials can also serve as adhesives. Here a fusible thermoplastic adhesive is melted in an electrical heating gun. Central picture at left: Bonding with silicone rubber (building of a frameless aquarium). Below left: Welding thermoplastic sheet with a domestic apparatus.

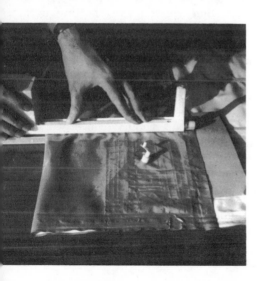

13.
Thermoplasts for Household, Leisure and Craft Work

Your choice of thermoplastic materials for do-it-yourself jobs is widespread. It extends from thin self-adhesive sheeting over thick covering materials with a smooth or engraved surface to profiles, such as curtain rails and edging profiles for chipboard panels, tubings and foam panels.

SELF-ADHESIVE SHEETING OF ALL KINDS

There can be few do-it-yourself fans who have never worked with self-adhesive sheeting, so that a detailed discussion is unnecessary. Nevertheless, a few hints may be useful, which can make your work easier and lead to better results.

All self-adhesive sheeting is equipped with so-called tacky glues, which provide an extremely good adhesion to the underground if its surfaces are dry and, what is most important, solid and free of dust.

Any surfaces that are not sufficiently stable, such as old plaster, should be treated with a primer that penetrates deeply into the plaster before the sheet is applied.

Due to the thinness and flexibility of self-adhesive sheets, any unevenness underneath will be visible, so that only smooth surfaces will lead to good results. Covering polystyrene foam with self-adhesive plastic sheets is a tricky job, as you cannot achieve a lasting adhesion. But you can overcome this trouble by giving the foam a coat of emulsion paint that is allowed to dry thoroughly before the sheet is applied. Even high-quality sheeting will inevitably turn slightly harder and also shrink a little over the years. Therefore, it is advisable to overlap when covering a piece of wood all around or when glueing the sheet on a cardboard drum to convert it into a paper basket. Allow both ends of the sheet to overlap about 1 in. (2.5 cm.) and try not to place the joint close to an edge. Otherwise, it might happen that the joint will open one day when the sheet is shrinking.

Sometimes, you may have some trouble when covering an object because the sheet is warping, even though the surface you are covering is entirely even and smooth. The reason is quite simple. You have stretched the soft and rather elastic sheeting unevenly by pulling more strongly on one side, so that this side becomes longer. Such a mistake cannot be cured once it has happened, but you can avoid it. However, there is an expedient: you can cut the stretched part off and add a new piece that overlaps the remaining part of the first. Too much pulling is also detrimental, even if it is done evenly over the full width of the sheet and warping is avoided. This is because the sheet has a certain tendency to return to its original state and will slowly contract in time, so that there may be a gap one day if the overlap was not wide enough or if the ends were butt-jointed. Even the lateral edges of coated panels may become exposed, because the sheeting will also contract crosswise by some fractions of an inch. Special care must be taken when lining the inside of cylindrical objects. The sheet must never be stretched by pulling, and you should allow a generous overlap at the joint where both ends meet. If the pattern allows you to do so, the inner lining can also be assembled from several strips that overlap each other.

If the overlap is big enough, it will allow for shrinkage, so that at worst there will only be a tacky dark stripe that can be easily wiped off with a rag moistened with alcohol.

Stretched sheeting will contract again in time

Sometimes do-it-yourself workers complain of air bubbles that are entrapped between the sheet and the covered surfaces. This mistake can be easily avoided if the protective paper on the rear side of the sheet is only removed step by step, especially if a long strip is applied. Just roll the sheet onto the surface, starting from one of the narrow ends. Small bubbles may be removed by piercing them with a thin needle and smoothing the sheeting towards the fine hole.

Avoiding air
bubbles

To avoid any warping and bubbles, roll the sheet, precut to the required size and shape, on a piece of round wood or a cardboard tube, with the protective silicone paper on the outside. The paper is then unrolled about one turn only, and the free end of the sheet is now precisely placed on the surface to be coated. Smooth it into place with a soft rag, and slowly unroll the sheeting as more and more of the protective paper is removed. This technique prevents the formation of bubbles and, at the same time, ensures firm contact of the sheet with the ground material. Join the edges with a thick razor blade, like those used for little balsawood planes, or a sharp trimming knife with a replaceable blade, which you can buy in every do-it-yourself shop, or just a sharp pair of scissors.

Padded Sheeting

Besides an immense variety of self-adhesive decorating materials, which are available in plain colors and patterns, with velvet or fabric finish, as well as in mat or high-gloss polyester-finish wood effect and which can be used wherever the surface need not withstand high mechanical strain, you can also buy high-gloss and mat quilt sheeting. It can be self-adhesive or have a nonadhesive cellulose back, which is fixed to the plastic by quilting stitches. The latter type is glued in place with wallpaper glue (cellulose glue), while the plastic sheet cover overlaps at the edges and is glued with an emulsion glue. Any glue that may come out of the seams must be immediately removed with a wet sponge or rag. You can also hide the seams under a length of ornamental braid, which is also used with self-adhesive quilt sheeting.

Quilt-type plastics are mainly used for lining the walls of cloakrooms and doors of wardrobes, the lining of showcases and cabinets, as well as for facings around mirrors and linings of trunks and other containers for clothes and linen.

PLASTIC VENEERS AND DECORATIVE MOLDINGS

Plastic veneers, which are generally made from flexible PVC or, sometimes, polystyrene, are much more resistant than the thin self-adhesive decorating coverings. They are easy to apply and heat resistant up to 140° F (60° C). Such veneers also withstand moderate mechanical strain. Do-it-yourself workers use them for lining cupboards and other furniture and for facing outside surfaces, as well as for covering doors and walls made from cardboard-lined plaster panels or chipboard.

Such veneers are supplied in different wood-type finishes and in a great number of colors, in some cases specifically matching the colors and fittings of laminated sheet from the same manufacturer, and can be used together with the highly resistant panels. This helps to save money when building one's own furniture and doing other jobs, as thermoplastic veneers cost less than half as much as laminated sheet. Wherever only moderate mechanical, chemical and thermal strains have to be encountered, such plastic veneers are ideal, while heavy duty surfaces, such as tabletops and working panels, should be covered with laminated sheets. Some consist of flexible PVC that can be easily cut and bent, and some of these are lined with a layer of high-quality cellulose, which makes it easy to glue them in place. The wrong side is absorbent and allows you to use any commercial glue. White glue, which is normally used for glueing wood, permits very precise adjustment. If you want the veneer to stick instantly, you had better use a contact adhesive. Panels, the edges of which are also to be covered, can be coated in one process without any seams, as the veneer is pliant enough to be folded over by smoothing with a soft rag. This is done after the main surface is covered and the glue has turned hard. Now, give the protruding veneer a coat of glue and turn it around the edge. Once the covering is finished, any protruding parts of the veneer are trimmed away with a sharp knife, a ripping chisel or a small plane.

Cheaper than panels

You can also buy plastic veneers made from flexible PVC, which are 1/128-in. to 1/64-in. (0.2-mm. to 0.4-mm.) thick. Some are self-adhesive, while others must be glued in place with white glue or contact adhesive. The surface of these materials can either be entirely smooth, have a slight grain structure or, even, a fabric-type finish. Such veneers either imitate natural wood or have uniform colors.

Another material is made from impressed polystyrene film

or foil, which looks very much like natural wood due to the impressed grain texture. This material is fairly hard and rigid, so that it cannot be just bent around corners and edges. In order to bend it around an edge, heat must be applied by means of an electric hair dryer, a fan heater or an adjustable electric iron with a Teflon-coated bottom. Sufficient heating is most important and requires some experience.

If you try to bend the material cold, the bending line turns white, which results from an irreversible overstretching of the molecular structure of the material. Such a white line cannot be removed. Technicians call this phenomenon "white fracture." This type of veneer is glued in place with special contact adhesive or white glue. Its reverse side has a special lining, which provides an excellent glueing surface.

There also are some fairly hard plastic furniture veneers which are used for lining cupboards and which can be laid around edges simply by rubbing them. The rubbing produces enough heat to make the material pliant.

Plastic Coverings for Doors and Walls

You can give your room and cupboard doors a new look by lining them with 3/64-in. (1.3-mm.) thick decorative flexible-PVC covering material, which is fixed on the wooden surfaces with a special emulsion-type adhesive. Smaller surfaces, such as doors of cupboards, can be managed by do-it-yourself workers, but larger ones require quite a bit of skill and should be fixed in a press while the glue is drying. As you can unhinge a larger door, you can take it to a specialist if you wish to have it covered with such a material.

Heavy duty PVC wall coverings Recently, very attractive and hard-wearing PVC wall coverings have come on the market. They are fixed to the walls like wallpaper, and their wrong side consists of a supporting texture. As this material can breathe, you need not be afraid of condensation. It is especially recommended for children's playrooms, kitchens, halls and vestibules.

Decorative Moldings for Ceilings and Walls

If you want to embellish your home, you will find quite a

variety of plastic covering materials for walls and ceilings in the shape of vacuum-formed or injection-molded plaques or large reliefs that look like tiles, wood or masonry. They can be easily cut and are simply glued to the wall or ceiling. With deep-drawn (vacuum-formed) parts, a sharp pair of scissors or a sharp knife will do for cutting. Injection-molded plaques must be cut with a fret saw or piercing saw. The question whether or not to use such panels is a matter of taste, for they are imitations and can be recognized as plastic copies. Here you should use a simple rule of thumb, which helps in many similar questions: the more appropriate for the material involved and the more appropriate for the job you want to do, the better. This means that plastics should, as a matter of principle, remain discernible as plastics, and their special properties should be used to full advantage. Imitations are permissible where they are not immediately obvious by their look or feel—as, for instance, on the ceilings of some rooms.

PLASTIC FLOORINGS

PVC plastic floors have become quite popular for kitchens and bathrooms. You can buy PVC tiles, which generally are self-adhesive, or buy PVC floorings by the yard in different widths. These are not self-adhesive. There is a large scope of different qualities and thicknesses. The plastic material can more or less be filled with powderlike or fibrous fillers. The back of the flooring material may be coated with cork, felt or plastic foam or left uncoated. The total thickness varies between 1/16 in. and 5/32 in. (1.5 mm. and 4 mm.). As a rule, you can start from the principle that the harder the materials of the PVC flooring, the more fillers they contain.

With very few exceptions (for instance, indoor-outdoor carpeting, which is waterproof, weather resistant and suitable for bathrooms and kitchens, where it may lie loose on the floor), all PVC flooring materials have to be glued onto the substrata, which must have been previously leveled with filler. The smoothing of any unevenness is indispensable in order to avoid strain at the edges of holes or bumps, which will quickly ruin the flooring material. PVC flooring materials without adhesive coating are generally fixed with light-colored plastic or neoprene contact adhesives.

Hard tiles, with a high percentage of filler, are fixed with   Fixing
bituminous adhesives. They can be easily cut if they are slightly

heated with a fan heater or in an adjustable electric oven. Flooring materials with a special coating on the wrong side call for special adhesives that harmonize with the backing material (generally, emulsion or synthetic resin glues). Always store the flooring material in a warm room at 68° F (20° C) overnight to make it flexible.

Preparations

Before the flooring material is put in place, the floor underneath must be cleaned of dust and any remnants of grease (wax floor polish). The latter is easily achieved by scrubbing the floor with soda lye. Then, rinse with clear water and allow the floor to dry thoroughly.

After the single widths of PVC flooring have been cut to their correct size and shape, they are laid on the floor. Start at one of the long walls of the room and fold back lengthwise one-half of the width of PVC flooring, which is laid down next to this wall. Then, apply adhesive to the strip of floor that is now lying bare and fold the flooring material back on the floor in a sliding movement against the wall. This technique is better than just turning the PVC back because it helps avoid the formation of bubbles as the material is unrolled.

Then, apply adhesive to the remaining area under the material, but leave a 4-in. (10-cm.) wide strip next to the join with the next width of PVC free from adhesive. Carefully slide the flooring material over this surface. The next width should overlap the first one by ⅝ in. to ¾ in. (15mm. to 20 mm.). It is fixed in place in the same way. Press the flooring down with sandbags placed on a smooth sheet of wood, which disperses the pressure evenly over the whole surface.

Once the whole floor is covered, the overlaps are cut through along a steel rule with a sharp knife. Open the overlaps and remove the loose strips before you now apply a thin coat of glue to the floor under the loose edges of the flooring widths. If you hold your knife exactly vertical when cutting through the overlaps, the free ends of the neighboring widths will neatly come together and form a seamless join.

Carefully roll the join to press the flooring material well down. Now it shows if too much adhesive was applied, as it will squeeze through the join and soil the flooring material. If this happens, it must be quickly removed with alcohol or petrol. Finally, place a board on the seam and weight it with some sandbags to hold the joins well down until the glue has turned hard. Finally, nail quarter-round stripwood against the skirting boards to achieve a neat junction at the skirting, too.

THERMOPLASTIC SHEETINGS

These coverings have become nearly indispensable helpers to housewives, do-it-yourselfers, model makers and campers. They provide a reliable and hygienic packing material for fresh or frozen food. They are welcome wherever there is a need for waterproof packaging or to exclude dust and dirt. As you may see from the items produced by the plastics industry, the use of such substances is by no means restricted to the packing of food and other items. Other popular goods made from thermoplastic sheeting include shopping bags and protective coverings for tools and sports equipment, fitted covers for office machines and photographic equipment, protective coverings for the seats of motor cars, rainwear and even inflatable furniture.

A do-it-yourself enthusiast need not be frightened of working with plastic sheeting. He can make many useful things, such as a cover for his enlarger in his home photographic laboratory, protective covers for such delicate items as the wings of model aircraft or fuselages that are never clean enough to be put into the trunk of the car, because of remains of oil and model engine fuel in the tank, or maybe a practical bag with many separate compartments for all the bits and pieces required on a camping tour or boat trip.

Homemade covers

Some people who have tried to make such things from plastics will possibly object to such suggestions, alleging that all these nice plans will finally fail because amateurs do not have the means to join plastic sheeting properly.

There are, however, various special bonding agents and suitable contact adhesives, but they all require a very skilled hand. Their solvent content makes the material swell very quickly if the adhesive is not applied very cautiously and sparingly, so that the sheet creases and produces irremovable wrinkles that look rather ugly. For straight joins, you may use fabric-reinforced waterproof self-adhesive tape, which, by the way, proves to be most helpful for shortening plastic rainwear for children. It can be easily removed when the sleeves have to be let out again as the child becomes a bit taller.

Sewing plastic sheeting is never really satisfactory unless the material is of the fabric-reinforced type. Otherwise, it will easily break at the seams. There is another drawback to sewing, i.e., the seams are neither watertight nor hermetically sealed.

Welding the plastics is the answer to all these problems, and many amateurs may say, "If only we could." In fact, you can! The only tool required for this purpose can be found in any modern household: an electric iron with heat control.

Welding plastic sheeting with an electric iron is by no means a makeshift; it provides perfect results, with little expenditure of labor. The sketches on the opposite page will show you the best way to go about it:

Welding with an Electric Iron

Cover your working table with a strong, smooth and, if possible—to avoid discoloration—white sheet of thick cardboard, which must not have a plastic finish. The heat insulation provided by the cardboard layer helps to create an even heating of the plastic, which is essential for a good seam. Set the temperature of the iron for synthetics and wait till the little control light goes out for the first time, which indicates that the iron has reached the right temperature. To prevent the sheeting from sticking to the hot base of the iron or melting, which would result from direct contact, a sheet of silicone paper is placed over the sheeting.

Silicone paper as ironing cloth Silicone paper is used, for instance, to cover the sticky wrong side of self-adhesive materials and should not be thrown away. Another strip of this paper is placed between the cardboard base and the two layers of plastic sheeting to be welded. Always make sure that the glossy self-releasing side of the paper faces the plastic.

Waterproof and air-tight seam The plastics should overlap by about ¾ in. (2 cm.) for a good seam. Move the iron slowly over the upper layer of silicone paper and apply moderate pressure. The heat of the iron will

HOW TO WELD PLASTIC SHEETING WITH AN ELECTRIC IRON AND SILICONE PAPER

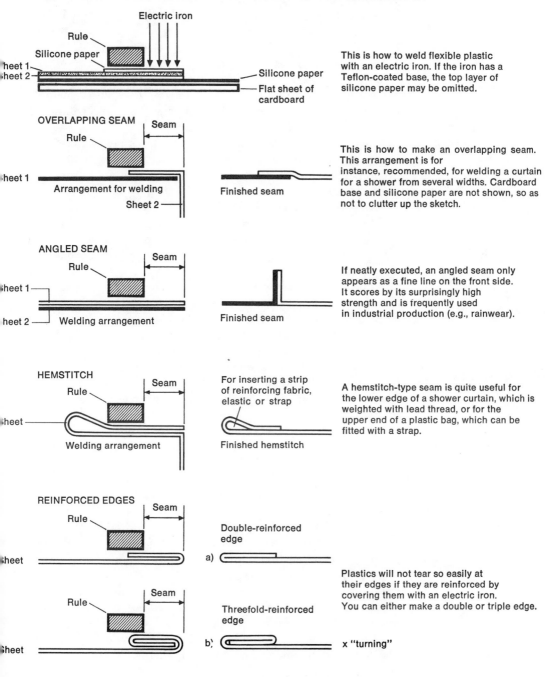

Electric iron

Rule
Silicone paper
Sheet 1
Sheet 2
Silicone paper
Flat sheet of cardboard

This is how to weld flexible plastic with an electric iron. If the iron has a Teflon-coated base, the top layer of silicone paper may be omitted.

OVERLAPPING SEAM

Seam
Rule
Sheet 1
Arrangement for welding
Sheet 2
Finished seam

This is how to make an overlapping seam. This arrangement is for instance, recommended, for welding a curtain for a shower from several widths. Cardboard base and silicone paper are not shown, so as not to clutter up the sketch.

ANGLED SEAM

Seam
Rule
Sheet 1
Sheet 2
Welding arrangement
Finished seam

If neatly executed, an angled seam only appears as a fine line on the front side. It scores by its surprisingly high strength and is frequently used in industrial production (e.g., rainwear).

HEMSTITCH

Seam
Rule
Sheet
Welding arrangement

For inserting a strip of reinforcing fabric, elastic or strap

Finished hemstitch

A hemstitch-type seam is quite useful for the lower edge of a shower curtain, which is weighted with lead thread, or for the upper end of a plastic bag, which can be fitted with a strap.

REINFORCED EDGES

Seam
Rule
Sheet

Double-reinforced edge

a)

Plastics will not tear so easily at their edges if they are reinforced by covering them with an electric iron. You can either make a double or triple edge.

Seam
Rule
Sheet

Threefold-reinforced edge

b)

x "turning"

penetrate through the paper and the overlapping sheets, which now soften, are automatically welded by the pressure, which is now slightly increased. Once cooled, the seam can be subjected to tensile stress. It will turn out to be not only very strong but also impervious to water and air.

Special care must be taken, however, to avoid any direct contact of the hot iron with the plastic, which would inevitably lead to a hole in no time at all. Make sure that the sheeting lies as smooth as possible and absolutely flat.

Temperature The exact temperature setting and the time required for proper welding require a certain degree of intuition and a few experiments with some waste strips of the substance you want to use. Both factors, temperature and welding time, depend on the type used and on its thickness.

By the way, even plastics of different colors can be welded if they are of the same type of plastic material. In case of doubt, an experiment is worth more than any theory and is not expensive if scraps are used.

For straight seams for shower curtains or light-proof curtains for your home darkroom, the iron is moved along a straight piece of wood. Curved seams are no problem at all, if you mark the required curve on the printed top side of the silicone paper with a soft pencil and then follow this line with the iron.

With a Teflon-coated iron If you are lucky enough to have an electric iron with a PTFE-(polytetrafluoroethylene-) coated sole plate (Teflon, Fluons), you can even dispense with silicone paper, as the softening plastic material will not stick to the coated sole of such an iron. Moreover, such an iron allows you to check the progress of your work, which reduces the risk of shifting which would result in a crooked seam.

If you use a Teflon-coated iron, however, the seam is not produced by a sliding movement of the iron but by pressing the iron down vertically step by step. Curved seams are more easily accomplished by using the curved sides of the iron or by joining such a seam together from a number of angled straight sections.

How to repair holes Even if you use an iron with a Teflon-coated surface, you must be careful not to touch the plastic with any other part of the hot iron because the edges of the sole plate are not Teflon-coated.

Even if your attention has wandered for a moment, and there is suddenly a hole in the plastic, you only have to cut away the crumpled edges of the hole and place a matching piece of plastic on top or underneath, which is then welded in place by pressing the heated iron onto it. This also works if you only have a normal iron and have to use silicone paper. If you work carefully, such a repair will be nearly invisible.

The same technique allows you to reinforce the edges of plastic sheeting by folding them over to form a hem, which is then welded with the iron. The edge will become even stronger if you fold again, so that three layers are then welded. With thicker sheets, it may be useful to weld a seam from one side, allow it to cool and then reweld it from the other side. With any kind of welding, you should always bear in mind that the seams must be allowed to cool down completely before they are exposed to mechanical stress. Otherwise, you will run the risk of the seam not holding or the plastic warping in its softened state. Such damage will not disappear on cooling and is permanent.

Folded edges can be reinforced further by lining them with strips of coarse gauze or fabric-reinforced plastic. Moreover, you can also make a "hem" and thread a string or elastic through.

Allow the seam to cool before exposure to stress

Welding with Special Equipment

You can also use special welding equipment that is normally used in the house for sealing plastic bags for deep-frozen food. Such a little machine is useful for many do-it-yourself jobs from plastic sheeting. But remember that all welding apparatus is not equally suitable for this purpose, as some of them are not fitted with heat control devices and are only designed for welding fairly thin polythene bags for frozen food.

As a matter of fact, all these little welding machines follow the same design. They look like a little desk and are equipped with an approximately 12-in. to 16-in. (30-cm. to 40-cm.) long electrically heated welding edge that is covered with a heat-resistant self-releasing Teflon film. The sheets that have to be welded are pressed together by a matching pad lever, which also actuates the switch for the heating wire when the lever is fully pressed down. A mechanism will switch it off again after a few seconds, also signalized by a control light that goes out and by the cessation of a slight humming noise.

The correct choice of welding temperature is partly a matter of instinct. Here again, it is best to make some tests with different heat settings, using pieces of waste material. The proper setting can be judged from the result. If the setting is correct, the heating edge will only mark a slightly mat line. Too low settings produce a dark line, while too much heat makes the sheeting wrinkle in the welding zone and may easily melt holes into it. The seam itself then looks twisted like a corkscrew.

But not only the appearance of the seam allows you to judge the proper welding heat. A tearing test is most informative, too, but you must, of course, allow the sheet to cool down before you test the strength of the seam. In a perfect welding, the sheeting will eventually be torn in the welding zone or next to it. If, however, the two layers peel off from each other without damage, the welding temperature probably was too low. The

How to weld an inflatable mattress
strength of such seams is astonishing. It even allows you to make inflatable mattresses and furniture from plastic sheeting. In a test made by the author, a large fully inflated air bag measuring 56 in. by 56 in. (1.30 m. by 1.30 m.), which was welded with a Bosch FG-16 welder, withstood the weight of a 198-lb. (90-kg.) person without any damage. In this case, however, the seams around the edges of the bag were reinforced by three single seams placed side by side.

There is a great disadvantage with all types of domestic welding machines: their welding edge is fairly short, and the hinge of the welding arm makes it very difficult to produce a seam in the center of large sheets. This problem can only be overcome by a special device. When welding a large inflatable mattress, for instance, the longitudinal seams that separate the single air chambers can be welded by folding the mattress. In order to prevent all the layers of sheeting from being welded together, a strip of cardboard must be placed between the two layers that are lying on the heating edge to be welded and the next two layers lying on top that are already welded. (See photograph on page 189 and diagram on page 199.)

But even this system has a tricky point, which is the kink where the sheets are folded. Here, you may run the risk of overheating. Such trouble can be avoided if you do not butt joint the ends of each section of the longitudinal seams but leave a small gap in between, which, at the same time, serves as an air passage between the single air chambers of the mattress.

It will be found advantageous to put such a welding

machine upside down if you have to make long seams that are welded step by step. Then, the apparatus lies flat on the lever, which is otherwise pressed down by hand. Now, you press the whole apparatus against this lever. Place the welder upside down for welding long seams

The main advantage of this trick is that the sheeting is nearly on a level with the worktable, so that it can be put more easily into the welder and does not shift so easily. Due to their straight welding edge, household welding machines only enable you to make straight seams, which appears to be a drawback for some jobs. Quite often, however, you will succeed in building up a curved seam from several straight seams (tangents). Again, use a piece of cardboard when making shorter seams to cover the welding edge wherever you do not want the seam to continue. A clever cut will quite often be a great help too, as you can see from rainwear that is made in large quantities from plastic sheeting. For easier working, straight seams are again preferred, and the sleeves are very often joined to the body part by a completely straight seam running from the armpit straight to the collar. In case of doubt, you can, of course, test the best cut on a paper model, which may then even serve as a pattern when cutting the sheeting. This method saves sheeting. If it finally turns out that a curved seam is unavoidable, you can still weld this section with an electric iron.

In order to prevent the rather slippery sheets from shifting, you can fix them with some small ends of double-sided self-adhesive tape, which are placed beside the actual welding zone. Only approximately ⅜-in. to ¾-in. (1-cm. to 2-cm.) long and ⅜-in. (1-cm.) wide strips of tape spaced every 12 in. to 16 in. (30 cm. to 40 cm.) will do, and they can be easily peeled off when no longer required. Possible remains of the tacky glue of the tape are removed with some alcohol or petrol.

The Third Way: Chemical Welding

Thick PVC sheets, which are, for instance, used for lining swimming pools or covering roofs, are often joined by a chemical process that is called chemical welding or swell-welding. Swell-welding

---

In the swell-welding process, the sheets to be joined overlap. The overlap is brushed with a welding agent, which slightly attacks the plastic and makes it swell. Then, both layers of sheeting are firmly pressed together, which causes the molecules of both softened plastic sheets to merge—a process that is very much like the knotting of the molecules in the thermic welding of thermoplastics described earlier. Once the liquid welding agent has volatilized, both sheets are firmly joined. The seam is both waterproof and airtight.

The success of this process depends very much on careful execution and proper cleaning of the plastics before welding. The overlapping seam must be pressed well down. In the do-it-yourself field, this method is of interest for pools only if the sheeting is not wide enough to cover the excavation in one piece. As these are once-only jobs, it is worth considering whether one would not do better to have this rather tricky job done by a specialist.

If you want to do the job yourself, you had better contact a professional and ask him for a demonstration of the process. An amateur will at best be able to weld about 40 in./min. (1 m./min.). Skilled professionals will cope with up to 120 in./min. (3 m./min.) if they work without special tools. With a chemical welding machine, which is used by the bigger roofing companies, a specialist can weld a seam of up to 22 yd. (20 m.) in only one minute, so that it appears to be simpler and safer to have the welding done by such a firm according to the design.

SWIMMING POOLS AND PONDS

You can even build a full-size swimming pool from PVC sheet. For this purpose a thickness between 20/1,000 in. to 60/1,000 in. (0.5 mm. to 1.5 mm.) is normally used; the sheet may be plain or fabric reinforced. There are many different systems of construction that are to a great extent suitable for do-it-yourself.

The excavation for the pool must be fitted with a metal frame

Large rectangular pools can be bought with completely prefabricated welded plastic linings hooked in an aluminum frame. As the plastic does not contribute to the structural strength of the pool but only provides the watertight sealing, an adequate substructure is required. The actual pool has to be built with bricks, which are then given a smooth coat of plaster. If you have to go that far, the question arises as to whether you had better not apply a glass fiber coating of three layers of glass mat and polyester, which costs no more than a prefabricated pool

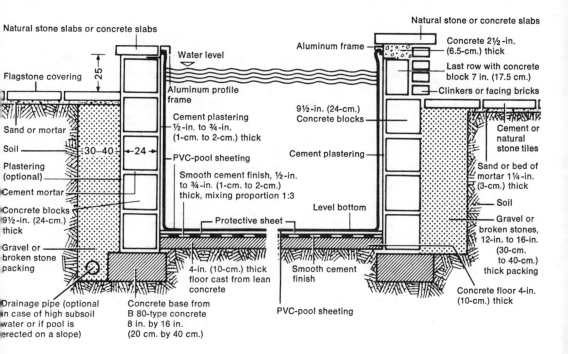

Natural stone slabs or concrete slabs

Flagstone covering

25

Water level

Natural stone or concrete slabs

Aluminum frame

Concrete 2½-in. (6.5-cm.) thick

Last row with concrete block 7 in. (17.5 cm.)

Clinkers or facing bricks

Aluminum profile frame

Sand or mortar

Soil

30–40

24

Plastering (optional)

Cement mortar

Concrete blocks 9½-in. (24-cm.) thick

Gravel or broken stone packing

Cement plastering ½-in. to ¾-in. (1-cm. to 2-cm.) thick

PVC-pool sheeting

Smooth cement finish, ½-in. to ¾-in. (1-cm. to 2-cm.) thick, mixing proportion 1:3

Protective sheet

4-in. (10-cm.) thick floor cast from lean concrete

Drainage pipe (optional in case of high subsoil water or if pool is erected on a slope)

Concrete base from B 80-type concrete 8 in. by 16 in. (20 cm. by 40 cm.)

9½-in. (24-cm.) Concrete blocks

Cement plastering

Level bottom

Smooth cement finish

PVC-pool sheeting

Cement or natural stone tiles

Sand or bed of mortar 1¼-in. (3-cm.) thick

Soil

Gravel or broken stones, 12-in. to 16-in. (30-cm. to 40-cm.) thick packing

Concrete floor 4-in. (10-cm.) thick

lining welded from plastic sheeting plus the aluminum frame required to fix it in the pool. The glass fiber coat costs about the same, but is much more resistant than plastic sheeting.

Besides a substructure built up from bricks, which has to take up the pressure of the soil, you can also have wooden or steel substructures; both allow you to install the pool on the ground or below ground level.

There also are round, oval, rectangular or oblong pools with round ends that have a supporting structure of steel profiles, which are lined with sheet steel, aluminum, plywood or glass fiber panels, plus an inner lining of prefabricated plastic sheet. Such pools can be raised above the ground or may be partly or even completely sunk into the ground. The base is generally covered with a thick layer of sand, on which the plastic is placed. Careful people even place an additional fabric-reinforced plastic over the sand. The sand must be free from sharp stones, which might pierce the plastic. There is also some danger from roots, which can easily grow through the lining.

Preparing the ground

If you want to dispense with supporting walls for a sunken pool, this can be done, providing the excavated walls are slanted at an angle of 45°. A thin, smooth, lean concrete finish is suitable if the soil does not contain clay that is firm enough to shape smooth and stable sidewalls. The bottom should again be covered with a layer of sand to protect the plastic.

Slanting walls instead of brick walls

Around the edges of the pool, the plastic should be reinforced with a fabric-backed strip, fixed with adhesive or, better still, chemically welded. This is best fixed with iron clamps anchored in the ground. Place stone slabs around the pool; they help to hold the plastic in its correct position. Here again, a thin layer of sand protects the plastic against damage from sharp edges of stone.

Ornamental Pools Are Child's Play

The same method can be applied for making ornamental pools, where slanting sidewalls are no problem at all. A 20/1000-in. (0.5-mm.) thick sheet is quite sufficient for this purpose. If the pool has a very complicated shape, you may even use two thinner sheets (1/128 in. to 3/256 in. [0.2 mm. to 0.3 mm.]) placed one on top of the other, which will more easily fit the contours of the pool.

Generally, a layer of sand will be quite sufficient to protect the plastic. For extra protection against roots, you can apply a layer of lean concrete, which must, however, be smoothed very carefully.

The right overflow

At first glance, it may appear rather difficult to make a proper overflow, which, just as with polyester pools, consists of a vertical earthenware pipe, the upper edge of which controls the water level. This pipe need not be connected to the sewage system and only requires a soakaway, as a high water level is generally due to rain, which normally seeps away in the ground. Place a coarse sieve in the tapered socket of the pipe, which retains any dirt or leaves floating on the water. Lay the sheeting over the pipe and pull it down into the socket with the sieve.

The sheet is then fixed with waterproof tape on the pipe. Wind the tape twice around the sheet, which is pulled tight on the pipe. Remove the sieve and cut the sheet out in this section, fix its edge inside and outside the socket with contact adhesive and put the sieve back in place.

About 4 in. (10 cm.) above the bottom of the pond, the sheeting is again fixed with a bandage of waterproof tape, while the joint between the pipe and the soil is smoothly filleted with sand under the plastic. Smooth the plastic in the excavation and distribute a generous layer of sand over the edge of the

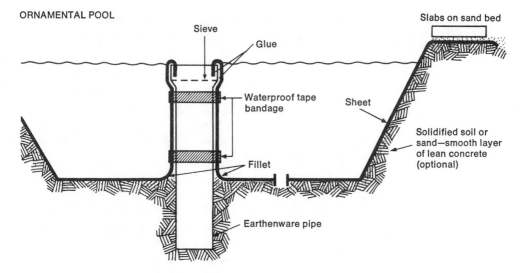

ORNAMENTAL POOL

Sieve

Glue

Slabs on sand bed

Waterproof tape
bandage

Sheet

Solidified soil or
sand—smooth layer
of lean concrete
(optional)

Fillet

Earthenware pipe

pool, to protect it against any sharp edges of the stone slabs that
are now arranged around the pond. Once the pond is filled with
water, you may put some sand on the bottom of the pool, which
makes it look more natural.

If you do not install an overflow pipe, which must be near
to one end of the pond to allow the sieve to be cleaned, you can
also put a fountain into the pool, which makes the water circulate
and enriches it with oxygen, which helps to control algae. In this
case, however, you must do without the layer of sand on the bottom of the pool in order not to run the risk of damaging the pump
of the fountain. Small underwater pumps with proper insulation
and special cable sheathing are not very expensive. Good pumps
are available at about $36 to $60 (£15 to £25). It is most
important that they are fully immersed in the water, as this
provides the required cooling for the motor. By the way, you can
also buy special plastics that enable you to keep fish in the pond.

A fountain
helps to
keep algae
at bay

HOT FORMING OF THERMOPLASTIC MATERIALS

One of the most important advantages of thermoplastics is
that they can be molded under moderate heat. Even though the
process takes place at fairly low temperatures, between 158° F
and 356° F (70° C and 80° C), compared with metal alloys, it
still requires quite complicated and expensive tools, so that the
widely used industrial forming methods, such as injection mold-

ing, extruding or blow molding, which supply us with cheap mass-produced articles are outside the do-it-yourself scope.

Only two hot-forming processes are exceptions to the rule, and even these are subject to certain reservations: these two forming techniques are deep drawing and bending. Both require the observation of a certain range of temperature, marked by two characteristic parameters: the softening point and the flow temperatures.

Softening point = freezing temperature

The softening point represents the moment when the thermoplastic material becomes flexible, at first elastic and, as the temperature increases, plastic. Flexible PVC, for instance, from which many plastic sheetings are made, has a softening point of −14° F (−10° C) and, therefore, is pliant and flexible at normal temperatures. If you put such a sheet into the deep-freezing compartment of your refrigerator, it will become rigid, because the temperature is now below the softening point. This is why this is also called freezing temperature. In other words, freezing temperature and softening point are the same. The two different names only result from opposite points of view.

The second value, the flow temperature, marks the temperature at which the material becomes pasty.

Between the softening point and the flow temperature, thermoplastics can be formed with little effort and, in fact, keep their new shape if they are cooled to below their freezing point. It is self-evident that this temperature must not be exceeded again when the part is used.

The greater the difference between the softening point and the flow temperature, the easier the hot forming of thermoplastics. The best molding temperature is just below the flow point. Then the least energy is required and the risks of white fracture are minimized if the material is not overstretched.

There are two groups of thermoplasts that differ in the temperature range in which they can be formed. The first one is formable between 158° F and 248° F (70° C and 120° C) and comprises the following materials: PVC (polyvinyl chloride), PVC/PVD copolymers (PVD = polyvinylidene chloride), celluloid (cellulose nitrate) and cellulose acetate. The second group is moldable within the range of 212° F to 320° F (100° C to 160° C) and comprises impact-resisting polystyrene, polymethyl methacrylate (acrylic glass) and cellulose esters (rigid packing films or foils).

Sources of Heat for Hot Forming

If you want to form thermoplastic materials, you must take care that the material is evenly heated. The forming temperature must be high enough, but, on the other hand, it must not reach the melting point or the temperature at which the material disintegrates.

All this may appear dreadfully complicated to the amateur, but the problem is not so difficult as it looks. In fact, there are suitable sources of heat in every house.

As all plastics burn and some are highly inflammable, naked flames are barred. But you can use the heating plates of an electric stove or cooker, an electric oven, preferably with thermostatic control or powerful infra-red lamps. If you have none of these, you can use an electric iron with or without a PTFE surface (refer to the section on Welding). Last, but not least, you can use hot liquids. In most cases, oil is better than water, as some plastics absorb water, which may cause bubbles or milky stains.

Hot forming can be dangerous for the amateur for two reasons. First, highly combustible thermoplasts, such as celluloid, may suddenly catch fire; they literally go up in smoke if they are heated too much. Therefore, you had better not use such materials but replace them by less flammable plastics. Cautious amateurs always test the flammability of the material by approaching a flame with a small sample about ½ in. by ½ in. (1 cm.²) held in a pair of pliers. Such a little sample will do for testing the flammability; bigger ones might become dangerous. Don't make any such test near other flammable substances; the best place for them is out of doors. If certain plastics represent a fire hazard or if there is a risk of burning one's hands, this applies no less to the sources of heat. Hot water and, especially, hot oil are most dangerous, chiefly because you are forced to work quickly in order to avoid loss of heat. To reduce the risks of burning your skin, you should always take pains to keep your

Check the inflammability first!

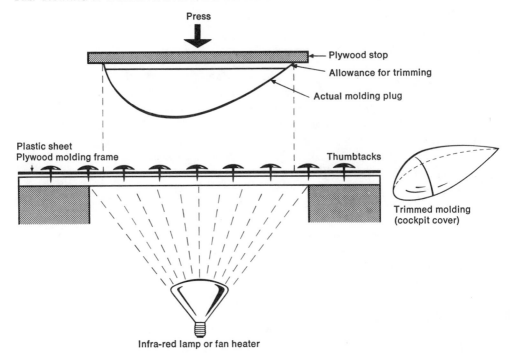

worktable clear of anything that you do not need at the moment. Heat-insulating gloves are advantageous too, because you will automatically try to catch a heated piece of plastic material that is slipping off your pliers. Finally, you must also manipulate the hot material to form it. The best tools for this job are padded clips, clamps or pliers with a large padded gripping area.

Deep Drawing

When forming thermoplastic materials by deep drawing, you start with a thin plastic, which is held in place with a clamping ring. For small parts, you can even use thumbtacks or drawing pins to secure the sheet on the molding frame. In factories and workshops, the sheet is generally held in a frame in which it is heated with infra-red lamps. After some time the molding plug with the male mold is raised from underneath and, at the same time, the chamber under the sheet is evacuated. Thus, the atmospheric pressure presses the softened and plastified sheet onto the mold.

After cooling, which can be speeded up by cold air, the cast is taken from the mold, and a new sheet is put into the molding frame for the next cycle, while the finished part is trimmed by cutting its rim away.

Deep drawing increases the surface of the original piece of plastic sheeting at the expense of its thickness, which is not uniform with such moldings and is very much reduced at the edges and curvatures. This is why the depth of the molding and the thickness of the original sheet material must somehow correspond with each other.

Molding depth and thickness of the sheet material must correspond

If this is neglected, the molding will break during the drawing process or get transparent patches due to overstretching of the molecular structure (white fracture). The stretching of the molecular chains, which is characteristic of this process, has a welcome secondary effect. The molded part is much stiffer than the original sheet material.

Due to the lack of suitable equipment, amateurs will generally have to confine their efforts to the molding of smaller parts, such as cockpit hoods for model aircraft, and only use sheeting with a maximum thickness of 1/32 in. (0.8 mm.), which allows a molding depth of approximately 1½ in. (40 mm.). The maximum thickness that can be mastered by amateurs with special tools and equipment is 40/1,000 in. (1 mm.).

Only small moldings for amateurs

The technique adopted by amateurs does, of course, differ from that used in industry. The amateur also uses a male master, which can be carved from wood or made from casting resin and must have the best possible finish. This plug-type mold is pressed through a cutout in a rigid plywood frame that has the exact outline of the molding plug but is slightly bigger than the contours of the plug require, as you have to allow a little space all around the plug for the sheet. Therefore, the cutout is as much bigger all around as the thickness of the sheet requires.

Male master

The plastic sheet is now fixed on this frame with thumbtacks or drawing pins—which must be spaced very closely—and is then heated with an infra-red lamp or fan heater until it becomes flexible and flaccid. For perfect results, heat the plastic not only in the region of the cutout but also around it, and also expose the plug to heat to prevent the sheet plastic from cooling down when it comes into contact with the molding plug.

Heat the sheet and the plug

Press the plug to the sheet and force it through the cutout in the plywood frame, the inner edges of which should be slightly rounded so that the sheet does not shear off at this edge.

Try to press the molding plug through the cutout in one run, so that the sheet need not be heated again.

The molding plug should be deeper than the molding depth that the part requires. At least allow the thickness of the plywood frame for extra depth, as the edges of the molding may be slightly wavy, so that this part should be trimmed off. If the molding plug is deeper than the molding should be, you will have excess material in this zone, which can be cut away so that the molding itself has a perfect edge.

It is also quite useful to fit the plug with a stopping plate that will lie flat on the plywood frame when the plug is fully pressed down. This guarantees that the molding will have exactly the shape it should have and that the plug is not pressed down in a tilted or oblique position.

Deep drawing requires some skill and experience, and an amateur will hardly succeed in his first attempt, so that he should not only have skilled hands but persistence, too.

Bending

Maximum thickness of 13/64 in. (5 mm.)

While a thickness of only 40/1,000 in. (1 mm.) is the maximum for deep-drawing thermoplastics with the rather primitive means available to most amateurs, sheet PVC or acrylic material (Plexiglas or Perspex, for instance) can be bent up to a total thickness of approximately 13/64 in. (5 mm.). Never try to make sharp-angled edges but always allow the material to form a slight curve, the radius of which should be at least twice the thickness of the material (for Plexiglas or Perspex). It will usually be found necessary to build a sort of bending tool into which the sheet material is put for bending and which must be fitted with suitable pressing pads. In order to protect the surfaces of the material, which easily get scratched, the supports and bending arms must be covered with soft glove-lining material if you want to bend acrylic glass. This covering will, at the same time, prevent the hot plastic material from sticking to the bending apparatus and also prevents marks in the softened material.

You may either heat the whole sheet or only heat the actual bending zone. Local heating can be achieved with infra-red lamps fitted with an opening mask of the desired shape, with electrical heating rods or radiators. You can even get flexible heating

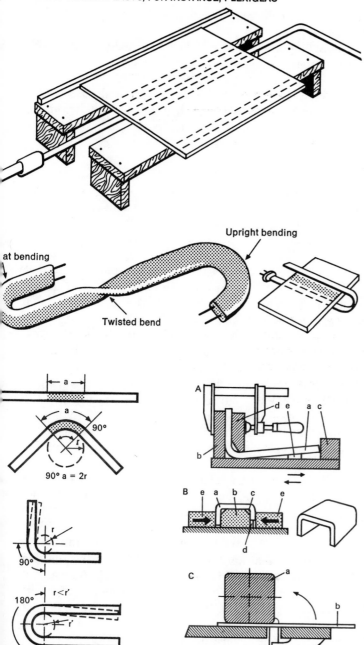

Upright bending

at bending

Twisted bend

a

a 90°

r

90° a = 2r

r

90°

180° r < r'

r'

A

d e a c

b

B e a b c e

d

C a

b

At the top:
The bending zone of a Plexiglas panel is heated with an electric heating rod.

In the middle:
Flexible heating rods allow a controlled heating of the bending zone, as they can be bent flat or upright and even twisted.

At the bottom:
If thermoplastic sheet material is locally heated, the width of the heated zone "a" must be at least as large as the development of the outside radius of the bend. The recovering of the material is balanced by bending the material more than the final angle requires (left). For exact bending, use a simple home-built bending tool, the surfaces of which are covered with glove-lining material.

A. SIMPLE TOOL FOR BENDING
a = Plexiglas panel
b = Stopping face (wood)
c = Stopping block
d = Retaining block
e = Loose support for more
   acute bending, which
   balances the recovering of
   the material when the
   part is taken out of the tool.

B. TOOL FOR BENDING A
   U PROFILE
a = Plexiglas or Perspex panel
b = Template
c = Tapered edges to prevent
   marks during the bending
   process
d = Spacing blocks for
   adjusting the
   length of the leg
e = Loose clamping blocks

C. DEVICE FOR EDGING
   SHEET PLASTICS
a = Wooden bending beam
b = Sheet Plexiglas or
   Perspex to be bent

These tools can also
be used for bending
other thermoplasts.

rods that are electrically heated and allow you to apply the heat exactly where it is required.

Heat a sufficiently wide strip of the material

It is most important that you heat the material in a sufficiently wide zone. When bending acrylic glass (Plexiglas or Perspex) at a right angle, you must heat a zone the width of which is at least five times the thickness of the sheet material. As a rule of thumb, the heated zone should be at least as wide, or even one and one-half times as wide, as the deflection of the outside bend of the finished part.

With very short legs, local heating of acrylic glass, such as Plexiglas or Perspex, may cause trouble: the bent part can warp when it is getting cool again.

Many amateurs are quite astonished when they realize that a part which was bent to a perfect right angle will no longer be right-angled if it is taken out of its mold. The legs of the angle have opened and now form an angle of 92° or 95°.

This effect occurs with all thermoplasts and is the more distinct the lower the temperature at which the part was bent. The reason for this phenomenon lies in the structure of the thermoplasts. As already discussed before (see pages 7–8), thermoplasts have long chainlike molecules that are entangled like the fibers of felt or may also lie parallel side by side. At normal temperatures, at which thermoplasts are rigid, the tangling of the molecules prevents them to a great extent from moving. Under the influence of heat, however, the molecules become mobile, while the molecular structure is slackening at the same time, so that the molecules can be easily shifted. When a thermoplastic material is formed under heat, its molecular chains are stretched. But they retain their tendency to return to their original arrangement. If you did not keep a bent thermoplastic part in the bending tool until it was cooled below the freezing point, but allowed it to cool freely instead, the plastic material would return to almost its original shape. By clamping the part into a jig, we apply force to the molecular chains and thus restrict their liberty to move. By cooling the part down again, the molecular forces that struggle against the bending or molding are, so to say, frozen or put into a strait jacket, but they are by no means eliminated. They still exist as internal stresses that are only wait-

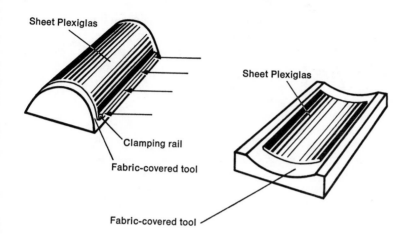

Sheet Plexiglas

Sheet Plexiglas

Clamping rail

Fabric-covered tool

Fabric-covered tool

ing for a chance to be set free, which may happen by mechanical strain, chemical influences or heat, of course. A small portion of these stresses, however, are not fully kept under control, and are the reason for a certain degree of recovery of the material, which is also called "memory."

You can use a trick to overcome this undesirable property of thermoplasts. You can outwit the material by bending it slightly sharper than is actually required.

A second way of controlling this effect is to have the right forming temperature. With rigid PVC, it is about 248° F (120° C); with cast Plexiglas it lies between 320° F and 356° F (160° C and 180° C); with Plexiglas xt, which is an extruded material, and Plexidur, a copolymer that consists of 70 percent of acrylonitrile and 30 percent methacrylate methyl ester, it lies between 284° F and 320° F (140° C and 160° C) and with polystyrene, the right temperature is within the range of 284° to 311° F (140° to 155° C).

A trick against the recovering: bend the material more sharply

Simple convex or concave parts from acrylic glass can be very easily molded in a wooden female mold or on a male mold, the surface of which is covered with glove-lining material. The sufficiently heated sheet material is only placed in or on the mold, where it is retained at its edges with a clamping device until it is cold again.

THERMOPLASTS FOR HOUSEHOLD, LEISURE AND CRAFT WORK   215

"MELTED PICTURES" FROM THERMOPLASTICS

Works of
art baked
in the oven

The low melting point of thermoplastics allows them to be turned into unconventional and most attractive works of art. The only requisites are a suitable source of heat (domestic oven with thermostat or a liquid-gas soldering torch), a piece of sheet metal, with a 3/16-in. (approximately 5-mm.) rim, silicone oil as a release agent and thermoplastic material in the form of powder, granulate, plastic foil chips and waste from sheet material or scraps from plastic articles.

MATERIAL		WORKING TEMPERATURE
Polythene	approximately	302° F (150° C)
Polystyrene		356–392° F (180–200° C)
Polyamide		482–527° F (250–275° C)

All these plastics are available in bright colors, which will actually gain in intensity and brightness by melting. To prevent the molten material from running off from the sheet metal base on which it is heated in the oven, use a metal plate with a 3/16-in. (approximately 5-mm.) high rim. The grainy powder or granulate is placed on this plate, which should first be treated with release agent.

The material is arranged in the desired shape in a thickness of approximately 5/32 in. to 3/16 in. (4 mm. to 5 mm.). Then the metal plate is put into the hot oven (see table above for correct temperature) where it must remain for five to ten minutes. Depending on the time the material is left in the oven, it will either produce a rough relief-type surface or an entirely smooth finish reminiscent of enamel. It can also be molded with a thin piece of wood while still hot.

Chilling with cold water makes the finished picture come off from its metal base more easily, but you must be careful not to scald your hands with steam or splashes of hot water. You may, of course, also do without a release agent so that the melted picture rémains on its metal base, which saves you a later mounting onto wood or chipboard. This is especially recommendable if you use polythene for your picture, as this material is most difficult to bond.

You may also arrange thermoplastic parts on a plain colored

surface or one consisting of fairly large pieces of different colors. If you bake this "collage," it will produce most attractive structures, as all bigger parts will get new contours in the melting process, the shape of which cannot be predicted.

While molten polythene produces opaque colors and is only translucent in very thin coats, the use of polystyrene allows you to make transparent or translucent pictures, bright enough to be reminiscent of stained glass. If you want to achieve an extra smooth finish, cover your picture during baking with a piece of sheet metal that has been treated with silicone oil or cutting solution. You will have no problem in removing your picture from the metal plates, as the plastic material shrinks considerably while cooling. Working with polystyrene requires good ventilation, as the material gives off poisonous vapors when heating. The same technique can also be used with polyamides, not all of which, however—due to their high melting and freezing temperature—can be melted in a domestic oven, which generally only produces a maximum temperature of 482° F (250° C). Polyamide is, as a rule, available only as a granulate or as cuttings. *Transparent pictures from polystyrene*

PAINTING WITH DISSOLVED THERMOPLASTICS

The technique of painting and decorating with dissolved colored thermoplastic granulates is slightly overshadowed by cold-curing enamel based on epoxy resin, which is easier and quicker to use. As a rule, one part by weight of granulate is dissolved in five parts of a solvent (trichlorethylene, toluol or benzene). These mixtures are kept in tightly sealed glass vessels —one for each color.

The liquid mixtures can be poured onto a metal plate treated with silicone oil. The dissolved colored plastics will slowly harden as the solvent evaporates. As long as the material is still flexible, you can even cut patterns from it or make reliefs, as it can also be modeled in this state. You may also allow the different colors to intermingle, as long as they are still liquid, to compose fascinating pictures with a great number of different colors. *Colorful pictures with intermingling colors*

Due to the long time required for hardening, which can take up to several weeks, this technique is not suitable for impatient characters. Above all, the solvents that are required for dissolving the plastic materials are by no means unobjectionable.

Toluol and benzene are even highly flammable. It is absolutely
essential to observe the following safety precautions: (*a*) ensure good ventilation and do not smoke in your workroom, and
High fire (*b*) bar open fires or other naked flames—including the pilot
hazard! lights of gas appliances.

14.
Thermoplastic Foams for Combined Structures

Expanded polystyrene is, no doubt, the main representative of this group of plastic materials. It can be easily recognized by its granular structure, which is visible at all cut edges. This material is offered in the shape of sheet material, blocks, profiles and moldings, such as foam shells for packing purposes and foam panels for coffered ceilings. Besides the standard type of foam, which is generally used for sheet foam, you can also buy a fire-resistant type at a higher price, which is advantageous for insulating your home and many other purposes. Standard polystyrene foam, as well as fire-resistant foam, has a specific gravity of 1.5 oz./cu. ft. to 2.5 oz./cu. ft. (15 kg./m.^{3} to 25 kg./m.^{3}). While sheets of these two types are made by cutting huge foam blocks with a grid of thin electrically heated steel wires, you can also get extruded sheet foam (see pages 9–10 for details on extruding). Extruded polystyrene foam has a tougher, entirely smooth surface and is also stronger due to its slightly higher

Expanded polystyrene sheets in three different qualities

weight of about 3.0 oz./cu. ft. (30 kg./m.³). You can buy all three types from a dealer in building supplies or a builders' merchant. The price varies according to size and thickness.

Ceiling panels and edging profiles, which are fixed to the wall to provide a good-looking joint with the foam panels under the ceiling, can be bought at do-it-yourself, paint and wallpaper shops and in large department stores.

You can improve the heat insulation of your home and save expensive fuel by either fixing polystyrene foam panels directly to the walls with a special adhesive or mounting them on a framework of roof battens attached to the wall. The framework is then faced with decorative panels, which, at the same time, provide protection for the foam and contribute the required compressive strength.

Especially the lighter types of polystyrene foam are extremely susceptible to point loads. Extruded polystyrene foam and PU foam panels (the latter do not belong to the group of thermoplastics) are more resistant. PU foam has a better insulating power than polystyrene foam. According to the current German standards for building materials, PU foam has a 12.5 percent higher insulating power than polystyrene foam of the same specific gravity and thickness. In laboratory tests, PU foam panels even proved to be 33 percent more effective. They both cost about the same.

How to reinforce polystyrene foam surfaces

If you do not want to install ornamental and protective facings, you can reinforce the foam surface before papering or painting the wall. This is done by giving the foam a coat of thick emulsion paint, to which 20 percent of synthetic resin binder is added. Then, a reinforcing fabric is embedded into this still wet first coat of paint. Allow twenty-four hours to dry, and then apply another coat of the same mixture with a sheepskin roller. If the surface is exposed to extreme loads, you can also apply a coat of synthetic resin plastering instead of a second coat of paint. The whole thing is much easier if you use sandwich panels consisting of a polystyrene foam core with outside panels of plaster and cardboard. Such panels are available in a thickness of 1¼ in. to 4 in. (30 mm. to 100 mm.) and can be used instead of plain poly-

Sandwich panels

styrene foam panels. They are mounted with a special mortar applied in several blobs on the back of the sandwich panels. The joints between the single elements, which measure 20 in. by 60 in. (50 cm. by 150 cm.) are smoothed with filler, into which strips of fabric are embedded to prevent the joints from cracking later on. The finished surface can be painted or papered.

HOMEMADE PLASTIC FURNITURE

Another interesting application of expanded polystyrene in combination with high-gloss solid polystyrene panels, which are generally used for vacuum forming, was discovered by the designer Gottfried Neuhaus of Frankfurt. He furnished a whole flat with homemade furniture and even used this material to decorate the walls, the ceilings and the floors.

Even though an amateur will not follow this example to its full extent, the system is interesting enough to be briefly described and to serve as a suggestion for do-it-yourselfers. The main point of this story is that the Neuhaus technique does not require special knowledge and that the few tools required can be bought for about $7 (£3) if you do not already have them in your tool box. Besides a long steel rule, which is normally used for papering, you will need nothing else but a keyhole saw, a multipurpose knife with exchangeable blades, a folding rule and a wax pencil for marking the foam and the 40/1,000-in. to 50/64-in. (1-mm. to 2-mm.) thick solid polystyrene panels. Small spatulas for applying the adhesive can be obtained from your local do-it-yourself shop free of charge, but can also be made of waste from the polystyrene facing panels.

No special knowledge required

All the furniture is built up from polystyrene foam blocks cut to shape, joined together and fixed to the walls. Any visible surfaces of the foam are then covered with 40/1,000-in. (1-mm.) thick polystyrene sheeting, which is available in different colors. You should use fire-resistant polystyrene foam or the stronger extruded polystyrene foam rather than the cheaper standard foam.

Cutting the foam to shape is no problem, even to novices, as this material can be most easily cut with a saw. Take care that you always move the saw vertically in order to produce straight and even cuts. If you are not very skilled in the use of a saw, you had better clamp two rigid lengths of plywood onto two wooden blocks of exactly the same thickness as the foam to be cut. Precisely adjusted with a T-square and secured with C-clamps, the two lengths act as a guide for the saw; put the foam through this jig so that the front edge of the wood coincides with the cutting line. This idea will of course help with straight cuts only. With curved lines—which are of special interest because of their elegant and attractive look and the wider scope of design—this method must fail. But you can produce perfectly true curves with an electrical hot-wire cutting device, which you can buy or even build yourself. It looks like a stationary fretsaw. Instead of the thin oscillating saw blade, such a cutter is fitted with a thin electrically heated wire that is tightly fitted between two insulated arms at a perfect right angle to the base on which the part to be cut is placed. The cutting wire receives electricity by means of a dry battery or a transformer with a potentiometer connected to the power source in order to adjust the heat of the wire. The foam is cut by the heat of the wire, which must not be too high. The foam is evenly moved forward against the wire with little pressure to produce a smooth and neat cut. The proper setting of the potentiometer can be easily found by some test cuts with scrap foam. It is correct if the cut edge is perfectly smooth and if the foam does not begin to run off due to excessive heat. As a rule of thumb, you should be able to touch the hot cutting wire for about a second without burning your fingers. The speed at which the foam is moved toward the cutting wire is another important factor, which you will soon learn to control after a few test cuts.

Cut curved lines with a hot-wire fretsaw

You can easily build your own hot-wire cutter from a model railway transformer, which must not have an overload release, an old fretsaw frame, which should be as long as possible, and a piece of chipboard to serve as a base. Thus, you get a useful tool for little money and a few hours of work.

If, however, you do not want to spend time and money on such a cutting device, you can still succeed with a little skill, as uneven cuts can be easily smoothed with a coat of foam adhesive.

Faced with polystyrene panels

The polystyrene facing panels can be broken over a straight edge, after the breaking line has been scored with a sharp modeling knife moved along a steel rule. Alternatively, the polystyrene panels may be lightly scored with a knife and torn like cardboard along the marked line.

The polystyrene facing renders the polystyrene foam surfaces, which are rather soft by nature, surprisingly compression proof. Convex and concave surfaces need not be fully supported with foam. It is quite sufficient to fix about 2-in. (5-cm.) thick formers, spaced at a distance of 4 in. (10 cm.) in place and then plank them with polystyrene panels cut to match the development of the curved areas. Try to make sure that the panels fit the formers snugly to achieve the greatest possible bonding area for an even distribution of any loads.

This type of design allows you to build ceiling-high shelves that are just fixed to the wall and do not require any additional reinforcements, as the author was astonished to see when he inspected an apartment equipped entirely with furniture of this type. The foam blocks are bonded to each other and to the walls with special paste adhesive for rigid foam, which can be bought from building wholesale suppliers or from builders' merchants. But you can also use a special contact adhesive for foams. Both types of adhesives are also suitable for bonding the facings to the foam.

Shelves up to the ceiling without additional reinforcements

Fixing and facing

The vertical joints of the facing panels are joined with polystyrene cement, which can be bought in modeling shops. Cut a V-shaped slit into the plastic nozzle of the tube. The width of this slit should be just as wide as the facing panels are thick. Put the slit over the edge of the panel you want to glue, and slide it along this edge while you lightly press in the tube.

This enables you to apply the adhesive evenly and so sparingly that none will be squeezed out of the joint to mar the high-gloss front of the polystyrene facing when the edges are joined. Cautious people also mask the facing at the edges of each joint with a strip of Sellotape, just to be on the safe side. Then, any possible trickles of adhesive can be quickly removed before they do any harm. Self-adhesive tape (Sellotape), by the way, is also most useful to hold the joints being bonded closely together for a neat and nearly invisible joint. The tape is peeled off after the glue has turned hard.

In addition to a minimum of practical skill, you will need quite a bit of talent as a designer and some feeling for shape and colors if you want to furnish your home with home-built furniture from foam and sheet polystyrene. Do not attempt to furnish a room at random, but make at least a rough plan showing the main dimensions and shapes. When designing curved surfaces, the natural flexibility of the material may be a great help. If you

Try to work to a plan

hold such a panel at its edges and then move your hands to the center and back again to the sides, plastic sheeting will automatically produce a graceful and evenly curved contour, which can be transferred to a strip of cardboard by an assistant who follows the contours of the plastic material with a pencil. Then, you can cut a template from the cardboard and use it to cut the foam panels and formers.

Drawers and handles

Drawers are assembled from 5/64-in. (2-mm.) thick sheet polystyrene. They are only reinforced by gussets that are glued into the inner corners of the drawers. You may also make drawers from ⅜-in. (10-mm.) thick foam parts that are faced inside and outside with 5/128-in. (1-mm.) thick sheet polystyrene.

You can use vacuum-formed parts to make handles for your drawers. Such parts can be bought from display shops and are fixed with polystyrene cement. Alternatively, you can cut flat ornamental handles from 5/64-in. (2-mm.) thick polystyrene sheets, which are fixed to the front of the drawer with a small piece of foam covered on all sides with sheet polystyrene, serving as a spacing block. Finally, you can also make the front about 1 in. (2.5 cm.) lower than the total height of the drawer, so that you can put your hand into the drawer to pull it out.

A face-lift for an old door

Once you have discovered just how attractive plastics can be, you will think of many more uses for the material. For example, vacuum-formed ribbed polystyrene panels can be used to make a dust-gathering door into a flush one. To mount the panels, first glue ⅜-in. to ¾-in. (1-cm. to 2-cm.) wide polystyrene strips (cut from waste) to the wood and, in addition, secure them with staples. Then fix the 20 in. by 36 in. (50 cm. by 90 cm.) ribbed panels to the plastic strips, which are placed on the very edge of the door. Another sure method: take approximately 3/64-in. (1-mm.) thick foam strips that are self-adhesive on both sides and glue them to the door near the edges; press the ribbed panels in place without using additional glue. The panels will stick permanently to the wood, and the door is not damaged, which is an important point if you are only a tenant. If need be, you can remove the panels with a sharp knife. The remainder of the self-adhesive foam strips can be removed with a little alcohol or petrol.

An illuminated ceiling from plastic bricks

Prefabricated thermoplastic parts can also be used for other most effective purposes. The prototype home designed by Neuhaus, for instance, has an illuminated ceiling made from translucent hollow plastic elements, which are the size of bricks and are bonded together at the fronts and sides. They are illuminated from the back with light from neon tubes, which do not heat the thermoplastic material and produce a pleasant diffuse light. If you want to use normal bulbs, you must take care that they are at a sufficient distance from the plastic and that only bulbs with low wattage are used. It is also advisable to provide for ventilation between the plastic ceiling and the actual ceiling of the room in order to avoid heat accumulation.

Blow-molded bulbous tubes with a length of 13 in. (33 cm.) can be assembled to form attractive suspended lamps. Glued Further possibilities together in the shape of an irregularly arranged bundle, these tubes make a most attractive pedestal for a low lounge table with a glass top, which does not obstruct the view of this attractive composition, as you can see in the color photograph in insert following page 116. Blow-molded high-density polythene spheres also make nice lamps if they are tied to each other with nylon thread or welded together. Welding is absolutely necessary, because polythene cannot be glued due to its waxlike surface. But you should not be frightened by the impressive word "welding," which, in fact, describes a most simple procedure: heat the blade of a knife and press the spheres against the knife where you want to weld them and allow the heat to act on the plastic material for a short time. Then, quickly remove the blade and press the spheres against each other. After a short cooling period, they are tightly welded together.

The same method can be used for joining the ends of PVC edging in a workmanlike manner. In this case, you press the ends of the edging against a hot blade of a knife or against the blade of a flat tool that can be mounted on an electric soldering iron. Once the plastic material has softened sufficiently, you press the ends of the edging against one another. The resulting flash is trimmed off with a sharp ripping chisel.

If the edging projects over the upper edge of the panel, the excess material is also cut off with a sharp ripping chisel. One-half of its blade, which is moved over the panel to cut the edging profile flush with the surface of the panel, must be covered with a strip of self-adhesive fabric tape to avoid scratches on the panel. Finally, the trimmed edging is smoothed with a piece of hard felt soaked with methylene chloride, which produces a perfectly smooth finish on the cut edge.

POLYSTYRENE FOAM FOR MODELING

Aeromodelers have been using polystyrene foam for quite a

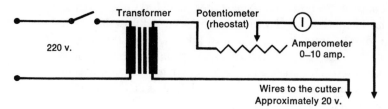

220 v.

Transformer

Potentiometer
(rheostat)

Amperometer
0–10 amp.

Wires to the cutter
Approximately 20 v.

WIRING DIAGRAM OF A HOT-WIRE CUTTING DEVICE

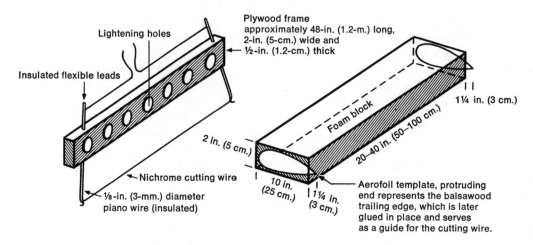

Lightening holes

Insulated flexible leads

Plywood frame
approximately 48-in. (1.2-m.) long,
2-in. (5-cm.) wide and
½-in. (1.2-cm.) thick

Foam block

1¼ in. (3 cm.)

2 in. (5 cm.)

20–40 in. (50–100 cm.)

Nichrome cutting wire

10 in.
(25 cm.)

1¼ in.
(3 cm.)

⅛-in. (3-mm.) diameter
piano wire (insulated)

Aerofoil template, protruding
end represents the balsawood
trailing edge, which is later
glued in place and serves
as a guide for the cutting wire.

TYPICAL TOOL FOR CUTTING POLYSTYRENE FOAM WINGS

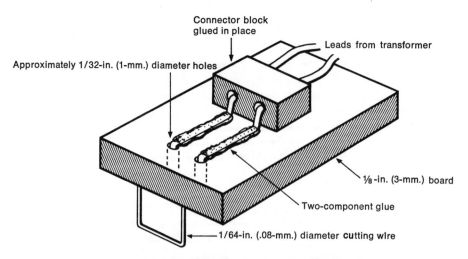

Connector block
glued in place

Leads from transformer

Approximately 1/32-in. (1-mm.) diameter holes

⅛-in. (3-mm.) board

Two-component glue

1/64-in. (.08-mm.) diameter cutting wire

SIMPLE GADGET FOR NOTCHING FOAM WINGS FOR REINFORCING SPARS

long time to build lightweight yet very strong wings, which are shaped with a hot wire and two templates glued on the sides of a polystyrene foam block roughly cut to shape. The cutting tool looks very much like a large violin bow, the string of which is replaced by a tightened heating wire that is guided over the edges of the rib templates at the end of the foam block and, thus, cuts the whole wing out of the polystyrene foam block. This technique allows the production of very precise aerofoils. The leading and trailing edges consist of strips of balsawood that are glued to the foam core. Both sides of the foam wing are sheathed with thin sheet balsawood (approximately 20/1,000 in. to 40/1,000 in. (0.5 mm. to 1.0 mm.), which is glued to the foam and must still be sanded, so that its thickness will be slightly less when the wing is finished. Thicker wings are sometimes reinforced by spars, which are glued into matching notches cut into the foam core. Dihedrals are achieved by glueing V-shaped braces, cut from hard balsawood or plywood, into slots in the wing halves. For additional strength, the joint between the two wing halves is coated with a layer of fine glass-fiber fabric and epoxy resin. All wooden parts, such as the balsawood leading and trailing edge, the balsawood or plywood braces, etc. are glued to the foam with white glue, while the sheathing is glued in place with a special contact adhesive for polystyrene foam, but you can also use white glue for this purpose.

The drawings on the opposite page show you how to build simple cutting tools for complete wings and for cutting notches into polystyrene foam wings for model aircraft.

Lightweight and strong wings

SANDWICH STRUCTURES FROM PVC FOAM PANELS AND GLASS FIBER

Besides PU foams, which belong to the group of duroplasts because of their resin base and which are supplied in liquid form as well as in molded panels, another type of foam, expanded PVC (which is a thermoplastic foam), is most suitable for making lightweight sandwich structures with a high bending strength.

PVC foam panels can be molded with little effort, if they are heated with an infra-red heater to a temperature of 212° F to 266° F (100° C to 130° C). This allows you to line even curved surfaces, such as boat shells, with foam. The curved panels are glued to the inner side of the glass fiber shell with thickened laminating resin. Very large panels should have approximately ⅛-in. (3-mm.) deep notches, which are scraped or cut into the foam every 2 in. (5 cm.) in order to allow any entrapped air to escape when the panels are glued in place. These channels guarantee a proper bonding of the foam panels to the glass fiber surface.

In order to keep the reaction of the styrene contained in the polyester resin as short as possible, to avoid damage to the foam, the surfaces that will come into contact with the polyester resin should receive a very thin coat of polyester filler, which should not completely hide the cellular structure of the foam, to ensure good bonding when the panels are glued on the glass fiber later on.

While the panels are glued in place, you should try to achieve an evenly dispersed pressure over the whole area. You do not need high pressure for glueing. Amateurs very often use small sandbags for pressing the panels down, but you can also use an inflatable mattress filled with water.

Proper relation of the thicknesses of the inner and outer shells

When building a boat with a sandwich structure, the outside shell, as a rule, is made 15 percent to 20 percent thicker than the inner laminate. According to the specifications for lifeboats issued by "Germanischer Lloyd" (the German safety council for ships), the foam core should be three times as thick as the total thickness of the laminate. In practice, you will find pleasure boats with even thicker foam cores.

If you are planning to build a sandwich-type structural part you should consult a specialist for the correct thickness and structure of the laminate and for the right thickness of the foam core. The strength of such a part can be calculated. It depends not only on the thickness and strength of the laminated shell and the foam core but is also, to a large extent, influenced by the quality of the bonding between the foam core and the surface structures. In order not to put too much strain on the bonding, it is necessary to provide a load-transmitting joint between the two skins when load-bearing elements have to be mounted inside the hull or if other local loads have to be taken up on one side of the sandwich structure. The easiest way of transmitting the load from one skin to the other is by laminating a large wooden block between the two skins. This block must have the same thickness as the foam core and will provide a structural connection of both skins that is resistant to shearing. Bolts that pass through both skins of the sandwich element are led through wooden inserts or through metal or plastic bushes, which should have a collar or flange on one side. Never forget to use a large washer on the opposite side in order to avoid stress concentrations.

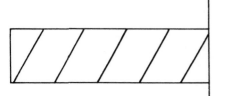

15.
How to Work Plastics— General Rules and Special Methods

Amateurs quite often face the need of cutting, drilling or machining plastics, which may be either thermoplastics or duroplastics. In most cases, normal wood- or metal-working tools will be suitable. But when using them, you should pay special attention to some rules resulting from the specific properties of plastics.

Poor thermal conductivity
    The most important point is that the poor thermal conductivity of all plastics must be taken into account. The thermal conductivity of plastics is only one one-hundredth to one one-thousandth of that of metal alloys, so there is a greater likelihood of heat accumulation on the working surface. When cutting, for instance, you must avoid excessive frictional heat, which causes thermoplastic materials to soften and become smeary. Sufficient cooling with diluted soluble oil or compressed air is, therefore, advantageous. Chucking of plastic parts may also cause some trouble. On the one hand, such parts often have a very smooth and slippery surface, and, on the other hand, the material is quite soft and has little impact strength, which easily

leads to visible marks on the surface. Such marks not only mar the finish of such a part; under stress, they may also become the starting point for a crack. For the same reason, you should also avoid angled cuts, but always give preference to rounded off corners when making cutouts in plastic panels, pipes and edging. Also, round off smooth transitions between the single sectors of tapered parts turned on a lathe. This technique is used by metal workers and has proved equally effective with plastics. The following pages give detailed hints for different working methods, with special sections dealing with mechanical shaping of thermoplastics, how to paint them and how to join such parts mechanically (by welding and with adhesives).

SPECIAL HINTS FOR THERMOPLASTICS

Sawing

All thermoplastics are easy to saw. As a rule, a normal fine-toothed wood saw, a whipsaw or a fretsaw will do. The latter should be fitted with a special blade for cutting plastics, though a medium-cut wood blade will do. Saw blades for cutting metal are less suitable, as their teeth are too fine and get quickly blocked by softening plastic material. Longer cuts can be made without trouble, if you use a table-type disc saw or fretsaw powered by an electric drill.

Saw blades for metal are too fine

Sheet material must, at all events, be evenly supported on a fairly large area, in order to avoid dangerous flutter, and should be firmly pressed down. Thin sheet material and acrylic glass, in general, are best cut with a very sharp saw disc or blade, the teeth of which are not side set.

The proper adjustment of the height of the saw disc is most important for a good cutting result. The teeth of the disc should not protrude more than a few millimeters (3/32 in. to 7/32 in.) over the surface of the material so that the saw cuts nearly tangentially. This will produce a very smooth surface. As only a very small part of the disc is moving, there is also less risk of choking the saw.

Cutting Plexiglas and other types of acrylic glass requires special care. Remove the protective paper after cutting, to avoid scratches. The saw disc should have as many teeth as its diameter measures in millimeters or twenty-five times the diameter in inches. Optimum cutting results are achieved at a cutting speed of approximately 9,842 ft./min. (3,000 m./min.), but this fairly high speed is rarely achieved with do-it-yourself tools, i.e., attachment of electric drills that rarely exceed 2,000 rpm under load and are only equipped with fairly small saw discs. All you can do is use the maximum speed; this also applies to most other thermoplastic materials.

Rate of feed and setting of the teeth

The ideal rate of feed depends on the special conditions given by the cutting tools and must be determined by some trial cuts. Acrylic glass with a greater thickness than ⅛ in. (3 mm.) must, at all events, be cooled with cutting solution or compressed air, according to the manufacturer's specifications. Cooling with cutting solution in most cases is quite a problem for do-it-yourselfers, as the electric drills generally used are not waterproof so splashing water may cause a short circuit. If only small cuts are required, you may dispense with special cooling or use a handsaw. If it turns out that cooling is absolutely necessary, it may be better to have the material sawn in a specialized workshop or buy the sheet material cut to size.

Polythene panels up to a thickness of 5/16 in. (8 mm.) can be cut with disks and saw blades with nonset teeth. Slightly set teeth are recommended for thicker sheet material of this type. Polypropylene, which easily becomes smeary, requires set teeth even for cutting thin sheet material. It is cut with the disc revolving at lower rpm and with increased feed.

Sheet low-density polythene, polystyrene and rigid PVC, as well as more flexible types of PVC, are best cut with slightly set teeth.

Blades with set teeth are indispensable for cutting curved contours with a band or piercing saw, to make sure that the blade does not choke.

In this context, reference should once more be made to the cutting of expanded polystyrene with a hot wire and the breaking of sheet polystyrene after scoring the breaking line with a sharp knife.

Universal tool: twist drills

Drilling

Twist drills, like those used for drilling metal, are a universal tool for drilling thermoplastics, too. Generally, twist drills with an angle of point of 110° and with the flutes of the drill at an angle of 30° against the longitudinal axis (of the drill) are suitable. Standard twist drills for steel meet these requirements if their diameter is more than 13/64 in. (5 mm.). Then, you need not reshape the drill by grinding. As the grinding of the point of a drill is a rather tricky job, amateurs will have to use standard twist drills for steel, even if the diameter of the hole is less than 13/64 in. (5 mm.). But you can also buy special drills for plastics. These are the best choice for drilling acrylic glass. With these special drills, the flutes are much steeper (angle of twist 12° to 16°), and the point is much more acute (60° to 90°). The stem of the drill should not be pointed.

Drilling may easily cause heat accumulation, and cooling by water, cutting solution or compressed air is required if the hole is to be deeper than 13/64 in. (5 mm.). The drill should also be lifted out of the hole frequently for removal of the grit and to help with cooling. Smooth cuttings indicate that you are working in the proper way when drilling acrylic glass and most other thermoplastic materials. By the way, if the drill is properly ground and sharp and if you ensure sufficient cooling, the walls of a hole drilled into clear acrylic glass will remain clear.

Besides twist drills, some special types of drills are very suitable. First, there are conical drills with two voluminous flutes that are parallel to the center line of the drill. For cutting large holes into sheet material, you can use a two-edged cutter with a central guide pin. Such cutters are adjustable to the required diameter and are also used for woodworking. They are frequently used for cutting holes with a diameter of ¾ in. to 1⅝ in. (20 mm. to 40 mm.). Even bigger holes are cut with a compasslike cutting instrument, which has two legs, one of them fitted with the cutting blade, while the other one acts as a centering pin.

Cone-type drills

Two-edged cutters and compass-type cutters

When drilling holes of small diameter, it may happen that the upper edge of the hole breaks out in the shape of a shell, which leads to a notching effect and may provoke a crack. This defect can be eliminated by careful burring and countersinking.

The best cutting speed (circumference of the drill times the speed of the drill in rpm) is between 164 ft./min. and 328 ft./min. (50 m./min. and 100 m./min.) if twist drills are used. Mind the units of measurement when calculating the cutting speed! For a 25/64-in. (10-mm.) diameter twist drill, for instance, the suitable range lies between 1,600 rpm and 3,200 rpm, according to the formula given above.

The right choice of speed of the drill

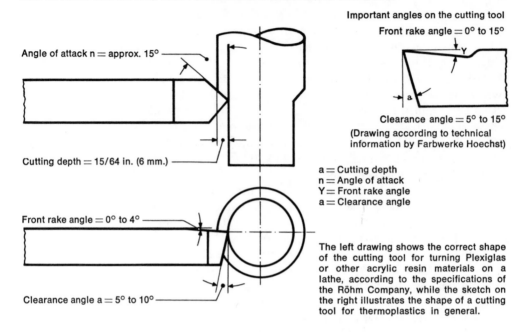

Important angles on the cutting tool
Front rake angle = 0° to 15°

Clearance angle = 5° to 15°

(Drawing according to technical information by Farbwerke Hoechst)

a = Cutting depth
n = Angle of attack
Y = Front rake angle
a = Clearance angle

Angle of attack n = approx. 15°

Cutting depth = 15/64 in. (6 mm.)

Front rake angle = 0° to 4°

Clearance angle a = 5° to 10°

The left drawing shows the correct shape of the cutting tool for turning Plexiglas or other acrylic resin materials on a lathe, according to the specifications of the Röhm Company, while the sketch on the right illustrates the shape of a cutting tool for thermoplastics in general.

Turning on a Lathe

Thermoplastics can be reasonably well shaped on a lathe with tools that are ground to meet the special requirements of these materials. With 820 ft./min. to 1,640 ft./min. (250 m./min. to 500 m./min.), the cutting speed of thermoplastics is, however, ten to twenty times that of steel.

Plexiglas and other acrylic resin materials ask for a rounded-off cutting tool with the greatest possible nose radius and a high cutting speed of 820 ft./min. (250 m./min.). Keep the feed low and the cutting depth very small in order to achieve a finish good enough for immediate polishing without further sanding.

With all other thermoplastics, you should use cutting speeds of up to 1,620 ft./min. (500 m./min.), a small rate of feed, but, if possible, a great cutting depth. As a rule, no cooling is required.

Milling and Engraving

The more rigid, harder types of thermoplastics can be milled and engraved with high-speed milling machines revolving at 20,000 rpm to 30,000 rpm, which are fitted with small cutting heads of the type used in dental laboratories. Even in this field, Plexiglas, Perspex and other acrylic resin materials are ideal.

Planing

The more flexible types of thermoplasts, such as PVC, low-density polythene and polypropylene can be trimmed with a normal plane for woodworking. The planed edges of PVC can be treated with a piece of hard felt soaked with methylene chloride to achieve a perfect finish.

Filing, Rasping, Scraping

Only fairly hard types of thermoplastics, such as acrylic resin, polycarbonates and polystyrene types, can be filed with normal files. With all the other thermoplastics, the fine cut of the file will be blocked very quickly. Woodworking rasps can be used or, better still, a Surform-cutting tool, which is fitted with hundreds of small cutting edges and is generally efficient. The cuttings pass through holes in the cutting surface and can hardly get stuck, and the tool can be cleaned with a turn of the hand. Surform-cutting blades are replaceable and are offered in many different grades of cut.

Files with normal fine cut are quickly blocked

Surfaces and edges that are shaped by rasping or with a Surform-cutting tool can be smoothed with a scraper.

Improvised scrapers are also very helpful for giving parts from acrylic resin the finishing touch. A scraper can be easily made by completely grinding the surface off an old, blunt triangular file.

Grinding the teeth off a high-speed steel saw blade turns it into a simple scraper.

Sanding and Polishing

Cut edges and the surfaces of the harder thermoplastics can be smoothed by sanding them with coarse to very fine wet and dry sandpaper. Acrylic glass, polycarbonate plastics, polystyrene and polystyrene foam, as well as expanded PVC, can be sanded quite well. Both types of foam will, however, only produce a finish that is similar to that of sanded balsawood.

If you are using a mechanical sander, be careful not to apply too much pressure and not to continue too long, so as to avoid ex-

Avoid excessive heat

cessive heat, which would cause the surface to become soft and smeary.

The cut edges of rigid PVC can be very well sanded with an oscillating grinding machine with dry 220- to 400-grade abrasive paper and are finally smoothed and polished with a piece of hard felt soaked in solvents. Flexible and waxlike thermoplastics, such as flexible PVC and polyethylene, cannot be sanded and may, at best, be smoothed with a scraper.

Plexiglas and other types of acrylic materials can be sanded by hand, with the sanding attachment of an electric drill or with an oscillating grinder or belt sanding machine.

Start with a relatively coarse grade of paper, depending on the original finish of the material, and then go on to finer and finer grades. But make sure that the sanding marks of the coarser paper used before are completely removed before you go on to the next grade. Start with dry paper, but the final sanding before polishing should be done with plenty of water and 400-grade paper. Excessive heat should again be avoided, as it could lead to stress and cause cracks.

Instead of wet and dry paper, you can also use steel wool (grade 0 or 00). Shaped scrapers or face-ground triangular scrapers can be used. Hollow-ground scrapers are unfit, because they produce undesirable marks. Finally, the surface has to be polished by hand with a soft cloth (felt, glove-lining material or a cotton buff), which is quite a tiring job, or with a buff wheel mounted on an electric drill. In both cases, polishing wax or cream must be applied evenly with a soft rag or a piece of wool to produce a high-gloss finish.

If you use an electric drill with a polishing pad, the whole procedure will be less tiring.

It is a prerequisite condition for good results that the surfaces to be polished have been carefully sanded, so that no sanding marks are left. Otherwise, the surface will indeed get a high-gloss finish but still show all the scratches and marks.

Apart from acrylic glass and polycarbonates, all thermoplastic materials can be polished only within certain limits. Here, again, you have to work with a soft polishing cloth and polishing wax, which will produce a silky sheen, providing the finish was more or less immaculate before you started to polish it. With luck, you may even achieve a high-gloss finish.

Tapping

Having drilled a matching tap hole, take a screw tap of the type used for cutting threads into metal and a suitable winding tool. As a matter of fact, threads in thermoplasts are rather risky because of their low strength. They are easily pulled out under load, so that industrial products made from thermoplastic materials are mostly fitted with threaded brass bushes that are embedded during the injection-molding process. With brittle types of rigid thermoplasts, there also is the risk of bursting when you try to tap the material. Threads cut into Plexidur withstand higher loads than threads cut into standard Plexiglas or other types of standard acrylic glass.

WORKING HINTS FOR DUROPLASTIC RESIN

Duroplastic resins are often mixed with very hard fillers. Quite often, the resin itself is hard, sometimes as hard and as brittle as glass. Both factors have to be considered when working. If you plan to work with duroplastic materials on a large scale, hard-metal-tipped saw blades and drills are the right choice. The susceptibility to notching, as well as the hardness and brittleness of these materials, must be taken into consideration when chucking or clamping a piece and when using any kind of tools. Otherwise, the material may get tension cracks, burst or break out. Sanding dust should be sucked off if possible, as it can cause severe irritation to eyes and mucous membranes.

Hard and brittle

Sawing and Cutting

Handymen generally encounter duroplastic materials in the shape of glass fiber or decorative laminates (Formica, Micarta, Resopal, Berite, Marlite, Ultrapas).

Glass fiber is best cut with electric fret or circular saws that are fitted with hard metal-tipped blades, the teeth of which are not at all, or only slightly, set. Professionals often cut glass fiber laminates with a silicone carbide abrasive wheel, which produces very smooth cuts and reduces the risk of delaminating, which is always imminent (i.e., the splitting of the laminate into its single layers). Such cutting wheels can be fitted with a cutting disc but are also supplied by some manufacturers as an accessory for the circular saw attachment of your electric drill.

Decorative laminates call for very high cutting speeds in

order to achieve smooth cuts without breaking off at the edges. Therefore, one should always use the high-speed gear of the electric drill driving the saw. The saw blade or disc must be sharp and its teeth should not be set. When the laminates are sawed, the decorative side of the panels should always be on the top. Work with a smooth and large support. The best thing to do is to press the panel down on the sawing table or workbench with a length of plywood to prevent flutter. The cutting line is marked with a wax pencil and can be easily wiped off later. The risk of the decorative top breaking off can be reduced if you place a length of transparent tape on the cutting line and smooth it down well. It is then cut, together with the decorative laminate.

You may get neat cuts, even with a hand saw, if you use saw blades with fine teeth that are not at all or only slightly set. Saw with short cuts from the decorative side. Quite good results can be achieved with a manual nibbling tool, which also proves most useful with many other plastics. It can be used for both straight and curved cuts.

Finally, there is the method of scoring a fine notch in the top side of the panel. Use a sharp tool, such as a ripping chisel, a pointed knife or a marking iron, and move it along a straight piece of plywood or a metal rule and repeat the scratching until the notch has reached half the thickness of the core of the panel. Then break the panel by moving it *upward* along a straight edge pressed down on the notched line.

Any other duroplastics, which are mostly in the form of molded parts, can be cut with a fretsaw, as required. Use fine saw blades, like those used for metal.

Drilling

Even drilling involves a certain risk of delaminating, and laminated duroplastics should rest on a block of wood during drilling. Sharp high-speed steel twist drills or even hard metal drills are best for this purpose. The point of the drill, which has an angle of about 120° with twist drills for metal, should be more acute and have an angle of 60° to 80°, which is also suitable for drilling decorative laminates. The latter are, however, best drilled with special drills for plastics similar to those recommended for Plexiglas and other acrylic materials (steeper and wider flutes than with drills for metal).

For cutting holes with a diameter of more than 25/32 in.

(20 mm.), it is better to use a two-edged cutter with a centering pin. For perfect cuts, it is best to work from both sides. For more than 2-in. (50-mm.) diameter holes, you had better use a compass-type cutting tool fitted with a hard metal tip. When making cutouts in decorative laminates, it is advisable to drill holes in each corner of the cutout and then make cuts from one hole to the next one and so on to produce a rectangular or polygonal cutout, but do not cut away the radius in the corners. The rounded corners help to avoid the stresses that can cause fissures due to the notch sensitivity of the material. The holes drilled into the corners should have the largest possible diameter, never less than 9/16 in. (14 mm.).

Filing, Sanding, Polishing

Glass fiber laminates, as well as blocks from casting resin, can be smoothed with sharp, not too coarse files. Rasps cause the edges to break out easily and may also cause delamination. Rough shaping of massive blocks can be done with Surform-cutting tools, which are, however, less suitable for laminates.

Cut edges of laminates can be smoothed by sanding them by hand or with a sander. Blocks are sanded in a similar way as acrylic glass (see pages 235–236). You may either sand them by hand, using a cork block covered with wet and dry abrasive paper, or use a rotary or oscillating mechanical sander.

Finish with wet and dry paper, after which the surfaces are treated with special polishing pastes suitable for the hardness of the material. You may, of course, also sand and polish the surfaces of the laminates. This process, however, calls for a sufficiently thick sealing coat of resin to prevent the reinforcing glass fibers from being laid bare. Surfaces that have become dull in the course of time can be retouched with polishing paste at any time. Decorative laminates have a very hard high-gloss or silky finish, which does not require any touching up. Once such a finish has dulled or has got scratches, you will have to buy a new panel, which may be glued on top of the old one, which should have been slightly sanded for better adhesion.

Decorative laminates used to be fixed with contact adhesives, which did not allow any adjustment once the panel was put in place. Therefore, it was standard practice to cut such laminates with some margin on both edges. The allowance of about 5/64 in. (2 mm.) had to be filed off later, for which Surform-cutting files were best suited. For finishing and chamfering the edges, a fine smooth file or 100-grade to 150-grade sanding paper can be used. When using sanding paper, you must carefully remove any loose grinding particles, which might otherwise cause scratches on the top side of the decorative laminate. However, you can avoid all this trouble and cut the decorative panel exactly to the required size, so that no extra trimming of the edges is required after fixing. A new type of contact adhesive with a jellylike consistency allows you to adjust the panel without trouble. Once correctly put in place, it is firmly pressed down, which makes the contact adhesive grip.

Tapping

Threads do not withstand high stress

Tapping is possible with quite a few duroplastic materials, but there is little point in it as the thread will not resist much stress. It tends to be torn out, or the whole part may burst if the plastic is fairly brittle. Standard screw taps for cutting threads into metal can also be used for duroplastics, but it is much better to use threaded bushes, which are embedded in the laminate, or, even better, tapped metal plates or flat iron bars, which disperse any loads more evenly. Another alternative: drill a hole all through and use a long bolt with nuts and sufficiently large washers to avoid stress concentrations.

HOW TO PAINT PLASTICS

For plain colors, the pigment is generally added to the liquid resin with thermoplastics and some duroplastic material or to the gelcoat in polyester or epoxy-bonded glass fiber laminates. Alternatively, laminates may have a colored film under a clear sealing coat, so that additional painting is not required. Pigments and surface finishes are generally very resistant, so that no touching up is needed.

Thermoplastics are susceptible to solvents

Thermoplastics are rather a delicate problem as far as painting is concerned as they are most susceptible to solvents and also have a very smooth surface, to which paint does not stick well. Slight roughening by sanding such surfaces is rather tricky, especially if the material is flexible. Polyolefin plastics, such as polythene and polypropylene, are hopeless anyway due to their extremely smooth surface.

Polystyrene: an exception

For these reasons, thermoplastics are usually left in their original color. Polystyrene sheet and foam, however, are exceptions. The latter can be painted with emulsion-type paints (for instance, when redecorating ceilings that are lined with polystyrene foam panels). Plastic model kits contain injection-molded parts from massive polystyrene, for which you can buy special types of lacquers which slightly attack the material and so adhere very well. They are available in high-gloss or mat colors. To make the paint stick even better, the surface should be cleaned with alcohol or lukewarm soapy water. Rinse carefully with clear water and allow to dry thoroughly. Small parts are best painted as long as they are still attached to the sprue and the runners connecting the parts during injection molding. This enables you to paint the parts completely without having to touch them after they are done. Leave the glueing areas free of paint. Larger parts, which have to be painted in one color, are painted all together after joining.

The lacquers may either be brushed on or sprayed. You can also paint small parts made from other thermoplastics, except polyolefin, with the same type of colors, which are brushed or sprayed on.

In order to achieve better adhesion of the paint, you can quickly wipe the surfaces to be painted with a little rag soaked with solvents, which will slightly roughen the surface chemically. Allow the solvents to volatilize completely before the paint is applied. Do not paint crash helmets with lacquer as the solvents in the lacquer may cause them to become brittle.

Duroplastics can be painted with any commercial enamel. Their surface, however, may carry a thin film of release agents from the molding process, which will hinder the proper adhesion of the paint. Washing the surfaces with solvents and slight sanding are advisable. Test the solvent on the wrong side so that even if you do make a mistake the part will not be spoiled. **Wash and sand duroplastics before painting**

The leading manufacturers of enamels offer special primers to be applied before painting duroplastics, which guarantee good adhesion of the enamel. Instead of standard synthetic resin enamels, you may also use highly resistant two-component polyurethane-based lacquers, which are also preferred for repainting glass fiber laminates. They produce better results than polyester air-drying lacquers. **Special primers**

DIFFERENT METHODS OF JOINING PLASTICS

The joining of plastic parts correctly and permanently with each other or with other materials quite often causes a severe headache to amateurs, because the job calls for different methods from those used for conventional materials.

Taps, Screws, Rivets

As mentioned earlier, in most cases, threads only resist minor loads and are easily pulled out. Bolting plastic parts together with long bolts passing through both parts to be joined is not the ideal answer either, as they quite often lead to point strain under load, which is not beneficial to the material. Sometimes self-tapping screws, similar to the sheet-metal screws you know from your car, are used for joints between plastic materials, if the joints only have to withstand fairly small loads and if the material is flexible enough to allow the fairly coarse thread of the screws to bite into the plastic material without bursting it.

Be careful with metal rivets Riveted joints are also possible. In most cases, spreading plastic rivets are preferred. Metal rivets may cause trouble, as they can notch the plastic material. This is the reason why fairly soft metals, such as aluminum, are preferred. Large washers under the head of the rivets and on the opposite side help to avoid notching and point loads.

Welding and Glueing

Large area joints are ideal Joints with a large contact area are ideal for plastics, as they allow even load distribution and help to avoid stress concentrations, if—and this is an important if—they are well made. In theory, welding and glueing meet these requirements. But only thermoplastics can be welded, and the handyman with neither the special tools (hot-air welding device) nor the indispensable experience at his disposal should stick to sheet. Apart from the welding of the joints of plastic edging and small polythene parts, i.e., joints that only have to withstand small loads, the welding of profiles and sheet material should be left to the professional.

Modern adhesives, which are to a great extent made from synthetic resins, however, enable the amateur to produce joints of extremely high strength, if the right type of adhesive is used for the materials to be bonded together and if the surfaces are thoroughly cleaned and roughened, and if—last, but not least—

the joint is designed in the best possible way to take the loads it has to withstand. The latter, as a rule, calls for sufficiently large bonding areas and an appropriate shape of the joint.

Glued joints may be subject to three different kinds of strain: (*a*) tensile load, (*b*) combined tensile and shearing loads and (*c*) peeling loads. Glued joints will generally not very well resist forces acting at right angles to the bonding surfaces (tensile loads), as the tensile strength of the adhesive in most cases is not as high as the tensile strength of the materials bonded together.

Peeling loads are most detrimental. They occur if one of the bonded parts curves under load and is lifted from the other part. This leads to stresses in a vertical direction to the bonding surfaces, which are concentrated in a very small area, i.e., the very edge where the part begins to peel off. The resistance of a glued joint against peeling forces depends on the elasticity of the adhesive and on the flexibility of the bonded materials. The stiffer the materials, the smaller the risk of peeling.

The fairly low resistance of glued joints against peeling forces is a fact we unconsciously make use of when we remove a Band-Aid or adhesive plaster from our skin.

Glued joints, however, prove to be quite resistant against combined tensile and shearing loads, which, for instance, occur with overlapping joints of two sheet metal parts that are exposed to tensile loads, especially because the strength of the joint can be considerably increased by increasing the bonding area. This allows you to transmit fairly high loads by such a glued joint.

<hr />

You must, however, bear in mind that simple overlappings are also subject to an additional bending load due to the different planes of the tensile forces, if the joint is exposed to tensile stresses. With extended overlappings, this may lead to peeling stresses at the ends of the overlapping parts (see sketch on the right).

<hr />

When planning constructive joints or repairs by bonding, the joints should be designed to keep peeling forces and tensile stresses under load as low as possible, to prevent nasty surprises.

There are five basic types of adhesives normally used by amateurs. Three of them harden by physical processes only, one of them either physically or by a chemical reaction and the fifth only by a chemical reaction. Emulsion-type glues consist of

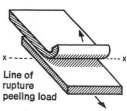

DIFFERENT TYPES OF STRAIN ON GLUED JOINTS

Tensile load

Combined tensile and shearing loads

Line of rupture peeling load

Bonded joints, low resistance to tensile loads

Five basic types of adhesives for the amateur

very fine particles of synthetic resin that are evenly dispersed in a solution. They are used for bonding porous materials together or nonporous materials to porous ones, such as decorative laminates to wood. The hardening is a rather slow process, effected by the evaporation of the water or its migration into the porous material, which allows plenty of time for adjustment of the parts to be bonded. A minimum temperature of 64.4° F (18° C) is required for hardening.

The second big group of adhesives are those containing solvent and swelling adhesives, which consist of synthetic resins dissolved in organic solvents. They are used for bonding nonplastic materials, like PVC and polystyrene, and also the so-called general purpose glues and modeling cements for hardwood and balsawood. Hardening is the result of a physical process, the volatilization of the solvent(s). Cooling and the transition from a plastic to a solid state cause fusible adhesives to turn hard. You can, for instance, buy them in the shape of thin films or foils, which are ideal for glueing decorative laminates onto wood or nonporous materials because they allow you to adjust the parts precisely. Such films are placed between the two surfaces you want to glue together and melt if heated. For this purpose, the amateur may again "abuse" the electric iron for ironing decorative laminates onto wood. The panel must be pressed down until the adhesive has cooled. Edging strips for chipboard panels can be bought ready-pasted with fusible glue, which makes them very easy to use. The edgings are ironed into place in no time, and there is no risk of glue being squeezed out of the joint, as frequently happens with other types of adhesives. Housewives also use fusible adhesives for mending textiles with glue-impregnated patches or fusible glueing powder, which allows them to iron a piece of fabric into place.

Contact adhesives are based on synthetic rubber (in most cases, neoprene) and are widely used for glueing decorative laminates onto wood and for glueing nonporous materials together. A thin coat is applied on both surfaces and must be allowed to dry until the surfaces are slightly tacky when touched but no longer stringy. The surfaces are then pressed together for a short time. Quite often, a single stroke with a hammer will do. Normal types of contact adhesives allow no further adjustment once the parts are put together, but this is possible with the new jelly-type contact adhesives. When bonding nonporous material, it is most important to allow sufficient drying before the parts are joined; once this is done, the solvents contained in the adhesive will no longer be able to volatilize, so that the bonding will never reach its full strength. Generally, contact adhesives become solid by a physical process, the volatilization of the solvents. But some types can also be mixed with a liquid hardener, and the resultant chemical reaction produces additional molecular links, which increases the bonding power and heat resistance. Adding hardener is highly recommended for covering chipboard panels with decorative laminates.

Reactive adhesives harden only by chemical reaction. They may epoxy, polyester, PU, vinyl acryl or cyanoacrylate shape of either one- or two-component systems. O nt reactive adhesives, which harden by the absorption of air moisture, are quite new and quite expensive. Above all, they are difficult to obtain for the amateur, because they have the fatal property of glueing fingers so tightly together that it needs a surgeon to cut them apart. As the adhesives harden in seconds, you should always wear polythene gloves or, at least, have suitable solvents at hand. In the United States, such bonding agents are classed as dangerous substances and are currently not available to amateurs.

Reactive adhesives may even bond your fingers together

If you happen to get hold of such a wonder glue, you should use it with the utmost care. Its bonding power is startling but frightening at the same time.

Reactive adhesives—either as one- or two-component systems —are used for bonding nonporous materials, such as duroplastics, metal, ceramics, stone, glass and even some thermoplastics (acrylic resin, polystyrene, ABS, rigid PVC, SAN [styrene-acrylnitrile copolymers]). Most reactive adhesives do not contain solvents.

Just to complete the chapter on adhesives, a further group of bonding agents must be mentioned that are known to nearly everybody but generally are only supplied as coatings on certain materials: permanently tacky adhesives.

Permanently tacky glues

Such adhesives consist of permanently tacky resins and serve as self-adhesive coatings for plasters, plastic sheeting, fabric tape and PVC film or foam pads or strips that are self-adhesive on both sides.

There is no hardening at all. Some types are even waterproof, like those used for making self-adhesive fabric tape, but most of them turn milky white if they come into contact with water and lose their adhesive power. Remnants of such adhesives can be removed with alcohol, white spirit, acetone, gasoline or other solvents.

No hardening!

Permanently tacky adhesives have proved most useful for all handymen (and women). They are available as double-faced foam pads for quick mounting on nonabsorbent surfaces (for instance, plastic articles on ceramic tiles) and for glueing polyolefin plastics, to which they stick moderately well and to which nothing else sticks at all. They can be used for temporary repairs to broken polythene buckets or for repairing polythene film or foil.

Bostik Pads The well-known Bostik Pads, incidentally, which also consist of foam, are coated with contact adhesives when they are used.

Bonding with solvents Bonding with solvents is an interesting method of joining thermoplastics. This process is related to swell-welding and is often used for joints requiring a very tight fit. The edges to be joined are only brushed with a fine brush, which is wetted with a matching solvent. The solvent is then allowed to attack the plastic material just enough to make it swell superficially before the parts are pressed against each other, so that the threadlike molecules will mesh. This method is of special interest to model makers and plastic kit fans, as it enables them to make very neat and practically invisible joints. Sometimes, some scrap plastic material is added to the solvent, which thus becomes similar to a solvent-type adhesive, which you can easily make yourself and adjust to any desirable viscosity and color. Only a few pieces of scrap material of the type of plastic you want to joint (for instance, the sprues and runners from a plastic kit) will do. There is not much point in using pure solvents, unless you want to join parts of the same plastic material.

WHAT TO GLUE WITH WHAT?*

Contact adhesives: Rez-n-Bond 1, Rez-n-Glue 123, Pattex, Uhu-kontakt, Fastbond 10, Terokal contact adhesive, Duro-plastic rubber, Bostik 1551, Bostik A 4, Bostik 233, Bostik Pads.

Two-component adhesives: Rez-n-Glue "EP," Uhu-plus, Uhu-plus-5-minutes, Uhu-end-feet 300, Stabilit-express (not suitable for polyethylene, flexible PVC, PTFE and polyamide), Araldite (different types), Duro-Epoxy-Glue, Bostik 778/779.

Resorcinol adhesives: Resarit, Aerodux.

Permanently tacky adhesives: Scotch Magic Tape, Rez-n-Glue WS-150, Sellotape, 3M-Scotchmount, Scotchtape.

* Only some of the brands of adhesives listed here are widely available in both the United States and Britain. Adhesives dealers and manufacturers can advise the home craftsman on comparable products.

General purpose glues: Rez-n-Glue 1 and 3, Rez-n-Glue AC series, UHU; Hard glues: Uhu-hart.

Dispersion glues (white glues): Rez-n-Glue AC series, Ponal, Mowicoll (for wood and foam), Bostik 708 (for plastic film or foil).

Special adhesive type A: for invisible joints of transparent and translucent acrylic materials: Rez-n-Bond, H-94, H-114, Permabond, Acrifix 90, Acrifix 92, Acrifix 93, Acrifix 96 (use Acrifix 95 primer for bonding Plexidur t and glueing Plexiglas and Perspex to other materials). Acrifix only hardens under the influence of ultraviolet and can, therefore, only be used for transparent parts. Other suitable adhesives: Tensol 2, Tensol 3, Tensol 6, and Reserit.

Special adhesive type B: Uhu-plast, Revell plastic cement, Faller PC-505/4.

Special adhesive type C: Uhu-por, Rez-n-Glue 213, Poron rigid foam glue, Dosbon 240, Bostik 597, Bostik 778/779, Assil-P, Assil-K, Ardal contact adhesive 50 or 80, RDC 3, Moltocoll rigid foam glue.

Special adhesive type D: VC-1, VC-2, VC-4, Duro-Plastic-Mender, Terokal-PVC glue, Bostik 475, Bostik 708, Bostik 785, Bostik 1475, UHU-PVC, Ovalit-K, Molit, Dytex, Tangit, Tip-Top repair set plastic A or plastic B.

Special adhesive type E: Rez-n-Glue 3, Bostik 1475 (special adhesive, primer and barrier for plasticizers, transparent, also suitable for bonding the materials specified in the following table to metal and other materials).

PU resin: Flexovoss K 6 TT (by Bondaglass-Voss Ltd.).

BONDING KEY

TYPE OF PLASTIC (general name)	SYMBOL	SUITABLE ADHESIVE	SUITABLE SOLVENT FOR BONDING
Acrylonitrile-butadiene-styrene	ABS	Contact adhesive, two-component adhesive and hard glue	No
Polyamide	PA	Resorcinol adhesive, two-component adhesive	Formic acid (85 percent concentration) plus 5 to 10 percent of PA material dissolved herein

TYPE OF PLASTIC (general name)	SYMBOL	SUITABLE ADHESIVE	SUITABLE SOLVENT FOR BONDING
Polyethylene (polythene)	PE	Permanent tacky adhesives, special adhesive type E	No
Polycarbonate	PC	Contact adhesive, two-component adhesive, hard glue, special adhesive type E	Ethylene chloride plus 10 percent maximum of PC material dissolved herein
Polymethyl methacrylate (acrylic resin)	PMMA	Special adhesive type A, contact adhesive, silicone rubber	Methylene chloride, ethylene dichloride, chloroform
Polypropylene	PP	Permanently tacky adhesives	No
Polystyrene (massive)	PS	Special adhesive type B, contact adhesive, two-component adhesive	Benzene, toluol, butyl acetate, methylene chloride (maybe with PS material dissolved herein)
Expanded polystyrene	(E) PS	Special adhesive type C, two-component adhesive, PU resin, epoxy resin	Foam is dissolved by solvents
Polyvinyl chloride (flexible types)	PVC	Special adhesives type D and E, dispersion glues for films or foils	Swell-welding agents, tetrahydrofurane
Polyvinyl chloride (rigid types)	PVC	Contact adhesive, two-component adhesives, polyester resin, PU resin plus G 4 primer, special adhesive type E	Methyl ethyl ketone, tetrahydrofurane, methylene chloride with PVC material dissolved herein
Polyvinyl chloride (expanded types)	PVC	Polyester, PU or epoxy resin, contact adhesive	No

BONDING KEY cont.

TYPE OF PLASTIC (general name)	SYMBOL	SUITABLE ADHESIVE	SUITABLE SOLVENT FOR BONDING
Polytetrafluoro-ethylene	PTFE	Cannot be bonded by amateurs at all!	
Cellulose acetate	CA	General purpose glue, hard glue, contact adhesive	Acetone
Epoxy resin	EP	Epoxy resin, two-component adhesives, PU resin, contact adhesive	Not possible because EP is a duroplast
Urea formalde-hyde	UR	Special adhesive type E, contact adhesive, two-component adhesives for decorative laminates; fusible sheet glue	Not possible; because UR is a duroplast
Melamine formaldehyde	MF	Contact adhesive, fusible sheet glue, two-component adhesive, special adhesive type E	Not possible because MF is a duroplast
Phenolic resin	PF	Contact adhesive, two-component adhesive, epoxy resin, special adhesive type E	Not possible because PF is a duroplast
Polyester resin (unsaturated)	UP	Fast-reacting polyester resin, epoxy resin, two-component adhesive PU resin, special adhesive type E	Not possible, because UP is a duroplast
Polyurethane resin (massive, inter-laced types)	PU(R)	PU resin, two-component adhesive, contact adhesive	Not possible, because PU is a duroplast or elastomer

TYPE OF PLASTIC (general name)	SYMBOL	SUITABLE ADHESIVE	SUITABLE SOLVENT FOR BONDING
Polyurethane (expanded types)	PU(R)	PU resin, polyester resin, epoxy resin, contact adhesive, dispersion glue	Not possible, because PU is a duroplast or elastomer
Silicone rubber	SI	Silicone rubber	Not possible, because SI is an elastomer

The bonding surfaces must be cleaned of grease, roughened and, if possible, also sanded

Careful preparation of the surfaces to be bonded together is a prerequisite for a firm and durable joint. They must be free from grease and roughened. Use alcohol to remove grease from thermoplastics that are susceptible to solvents. Wiping the surfaces quickly with acetone will help in most cases, and may lead to a slight roughening of the surface. Carbon tetrachloride is a very good solvent and cleaning agent for grease (caution: toxic vapors). Because of possible remnants of grease, additional sanding is recommended, even if the surfaces are already rough enough. Some commercial manufacturers of plastic items and some handymen rely only on sanding. Glass fiber casts must be sanded with special care, which also applies to compression-molded duroplastics, as any remnants of release agents on the surface—especially the silicone type—is fatal for subsequent bonding. Even if thermoplastics are not bonded with solvents or special adhesives containing a fairly high percentage of solvents (type B) it is recommended to roughen the surfaces by sanding to prepare them for the application of the bonding agent. This is of special importance when two-component adhesives are used.

When repairing household articles made from plastic materials (e.g., the casings of domestic appliances, tools, machines or model parts and toys), there is quite often the problem of how to achieve a lasting repair of thin-walled parts, especially as the repaired area will again be exposed to stress and the broken edges of the fragments only provide a very small bonding area. Therefore, it will be of little use just to bond these edges together.

Back the bonded cracks!

In such cases, you can reinforce the bonded parts by glueing a supporting bridge cut from the same material against the wrong side.

Allow the reinforcement to overlap the crack about 1 in. (2.5 cm.) to provide a bonding surface with enough strength to withstand stress. Embedding glass fiber fabric in a coat of adhesive is a most elegant way of reinforcing the wrong side of a repair, especially if two-component adhesives are used. For this form of backing, you must make sure that the surfaces are thoroughly roughened with coarse sanding paper to achieve the best possible adhesion.

Even though more and more impact-resisting plastics are being introduced, which, in addition, are reinforced with chopped-strand glass fiber, it still happens now or then that a plastic casing bursts or breaks. In such a case, the larger fragments can be assembled again. Bond the casing together, and reinforce it with a layer of woven glass fiber on the back. But the repaired part will not look very nice from the outside and, worse, may have sharp edges, involving risk to one's hands. Any gaps and cavities can be filled with two-component adhesive mixed with filings of the plastic material or some dry pigment to match the original color of the repaired part. When repairing burst thermoplastics, you may also dissolve the fragments that are too small to be glued in place again in solvents and allow them to thicken again, until you achieve a sort of filler paste. As this method is unlikely to provide enough material to fill all the cracks and cavities, because some of the fragments may have gotten lost, it is advisable to half fill all cavities with glue and to apply the homemade filler paste later on when the glue has turned hard. Allow the filler to harden thoroughly, then smooth it with very fine grain wet and dry paper. The old glassy surface can be restored with a coat of clear lacquer or, sometimes, by only wiping the surface with a rag soaked with solvents.

As you may see from the Bonding Key on pages 247–250, it is by no means unimportant to know what type of plastic you have to bond. If you cannot find a specification in the instructions, on the part itself or on the packing, you will have to make your own investigations. The following plastic identification table lists the major characteristics of the most important types of plastics and may be of help.

APPEARANCE/ COLOR	MANUFACTURED INTO	TOUCH	SCRATCHABLE WITH FINGERNAIL	HARDNESS	SOUND	BURNING CHARACTERISTICS AND SMELL	SPECIFIC GRAVITY	TYPE OF PLASTIC
Opaque, yellowish-brown, colored	Moldings, such as casings, boats and car bodies	Smooth	No	Hard	High	Burns with flickering bright yellow flame, smells like il-luminating gas and slightly like cinnamon	Approx. 1.06	Acryloni-trile-Butadiene-styrene (ABS)
Clear in the shape of films or foils, other-wise opaque	Films or foils, gear wheels, mechanical parts	Smooth	No	Flexible to rigid	Dull	Does not burn very well, once burning, melts and drops blue flame with yellow seam, smells like burnt horn	Approx. 1.02 to 1.14	Polyamide (PA)
Clear to turbid, black and trans-parent colors	Shopping bags, films or foils, buckets, toys, bottles and containers	Waxlike	Yes	Soft to medium	Dull	Little yellow flame with blue center, bright, smells like extinguished paraffin candle	Floats on water, 0.92+	Flexible poly-ethylene (PE)

Material	Density	Burning behavior		Hardness		Surface	Uses	Appearance
Rigid polyethylene (high-density polyethylene) (Hd-PE)	Floats on water, 0.92 maximum	Little yellow flame with blue center, smells like extinguished paraffin candle	Medium	Medium to hard	Yes	Waxlike	Pipings, beakers, insulation	Transparent, opaque or black and dark colors, respectively
Polycarbonate (PC)	Approx. 1.2, filled types up to 1.4	Only burns in a separate flame and stops burning outside a foreign flame, material smokes and becomes blistered, burning zone cools and produces a phenolic smell after some time	High	Very hard	No	Smooth	Insulations, unbreakable windows, lamps, heat resistant household items, housings for electrical machines	Crystal clear, transparent or opaque colors
Polymethylmetacrylate (PMMA)	Approx. 1.18	Bright flame, sizzling noise, melts, drops fruity smell	High	Hard to brittle	No	Smooth	Sign boards, name plates, window panes, moldings	Crystal clear, colorless, transparent or opaque colors

PLASTICS IDENTIFICATION TABLE cont.

APPEARANCE/ COLOR	MANUFACTURED INTO	TOUCH	SCRATCHABLE WITH FINGERNAIL	HARDNESS	SOUND	BURNING CHARACTERISTICS AND SMELL	SPECIFIC GRAVITY	TYPE OF PLASTIC
Opaque, gray	Moldings, housings, laboratory and kitchen utensils, liners for dishwashers	Waxlike	Hardly	Hard, tough	High	Yellow flame with blue core, material turns clear, black residues in the burning zone, resinous smell with additional smell of paraffin candle, material drops	Floats on water, approx. 0.91, filled types do not float, up to 1.3	Poly-propylene (PP)
Crystal clear, transparent and opaque colors	Plastic model kits, cheap plastic kitchen items and containers, toys, small fashionable furniture	Smooth	No	Very hard and brittle	High, nearly tinny	Burns with bright flame, drops, rising soot particles, sweet smell reminiscent of town gas or sometimes honey, modified types with additional smell of rubber or pepper	Approx. 1.05, modified types up to 1.08, foam from 0.01 upward	Poly-styrene (PS)

Material	Color	Uses	Surface	Breaks?	Consistency	Hardness	Burning behavior	Density
Flexible polyvinyl chloride (flexible PVC)	Transparent and opaque colors, also light colors	Foils, household goods, hoses, toys, inflatable toy animals and water wings, rainwear, fabric-reinforced covers, sportswear, floorings, edging profiles	Smooth	No, because flexible	Flabby, soft, pliant, leather-like and even flexible to rigid	Medium to high	Only burns in a separate flame, bright flame, sometimes formation of soot, stinging smell of hydrochloric acid, sometimes additional smells from plasticizers	1.2–1.35, filled types up to 1.6
Rigid polyvinylchloride (rigid PVC)	Translucent to opaque, often gray or different colors	Pipes, profiles, gutters, window frames, panels	Smooth	No	Tough to hard	High	Only burns in a separate flame, bright yellow flame, possibly with greenish line, smokes, burnt smell plus stinging smell of hydrochloric acid	1.2–1.5
Polytetrafluoroethylene (PTFE)	Opaque gray, black or brownish color, also mottle finish or opaque	Self-releasing coatings of pots, pans, electric irons and tools	Very smooth, nearly greasy	Slightly	Flexible		Nonflammable, does not char, separate flame gets blue-green seam, stinging smell only after extreme heating	2.0–2.3

PLASTICS IDENTIFICATION TABLE cont.

APPEARANCE/ COLOR	MANUFACTURED INTO	SCRATCHABLE WITH		HARDNESS	SOUND	BURNING CHARACTERISTICS AND SMELL	SPECIFIC GRAVITY	TYPE OF PLASTIC
		TOUCH	FINGERNAIL					
Transparent, crystal clear but also tinted all through in opaque colors	Cheap vacuum-formed parts, films or foils for modeling purposes (canopies) and packings	Smooth	Yes	Flexible, hornlike	Dull	Burns with yellowish green flame, sparking, drops, smells like vinegar (acetic acid) and burnt cellulose (paper)	1.3	Cellulose acetate (CA)
Crystal clear, yellowish transparent, also tinted all through, contents of glass fiber reinforcing possibly visible	Highly stressed glass fiber parts, laminated panels for electrical and electronic mounting plates, embeddings	Smooth	No	Hard	Dull	Difficult to ignite, little yellow flame, smokes, smell varies, depending on the hardener, between esterlike and stinging hornlike smell, after some time phenolic smell	Approx. 1.2, filled types up to 2.0	Epoxy resin (EP)

Color / appearance	Typical uses	Surface		Hardness		Burning behavior	Density	Name
Crystal clear, whitish but also tinted in opaque colors	Molded parts (casings for switches and machines), basic layers of decorative laminates, binders for molded wood and chipwood	Smooth	No	Very hard	Medium	Hardly flammable or only burning in a separate flame, makes separate flame turn yellow, carbonizes with white edge on the burning zone, stinging smell of ammonia or amines	Only filled types approx. 1.5	Urea formaldehyde (UF)
Crystal clear, white or tinted all through in any color	Sealing coats of decorative laminates and resin-bound molded wood, trays, unbreakable containers, insulating electrical parts	Smooth	No	Very hard	Medium to dull	Hardly flammable or burns only in a separate flame, turns the flame yellow, carbonizes with white seam on the burning edge, penetrating smell of fish	Only filled types approx. 2.0	Melamine formaldehyde (MF)

PLASTICS IDENTIFICATION TABLE cont.

| APPEARANCE/ COLOR | MANUFACTURED INTO | SCRATCHABLE WITH | | | SOUND | BURNING CHARACTERISTICS AND SMELL | SPECIFIC GRAVITY | TYPE OF PLASTIC |
		TOUCH	FINGERNAIL	HARDNESS				
Dark colors, mottle finish or sometimes uniform brown color	Handles and hand wheels, insulating parts, housings, gear wheels, bearings, heat-insulating handles on pots and pans	Smooth	No	Very often brittle	Dull	Difficult to ignite, burns mostly only in separate flame, then bright flame, smokes, suffocating smell of phenol and formaldehyde, additional smells by fillers	Pure resin approx. 1.3, filled types 1.8–2.0	Phenolic resin (PF)
Crystal clear, yellowish transparent or tinted in opaque colors, glass fiber reinforcement possibly visible at broken edges	Large moldings, such as boats, car bodies, furniture, corrugated panels, embedding of decorative objects and electronic circuits, filler, coatings	Smooth, glass fiber reinforcing possibly perceptible on the rear side	No	Hard	Dull	Bright yellow flame, not filled casting resins smoke and soften when burning, filled types crackle and carbonize, glass fiber reinforcement is left	Pure resin 1.1–1.2 filled types up to 2.9	Unsaturated polyester resin (UP)

Polyurethane (PU)	Pure resin approx. 1.2, filled types up to 2.0, foams from 0.01 upward	Difficult to ignite, softens and drops, burns with bright yellow flame, smells stinging, provokes coughing	Soft, flexible to brittle, hard	Dull	No	Smooth	Lacquers castings, furniture (structural foam), insulations against heat, cold and noise (foam), coatings, moldings, shiny leathercloth with crushed finish	Mostly dark opaque colors and shades that are resistant to turning yellow, sometimes yellowish transparent, rarely crystal clear
Silicone rubber (SI)	Approx. 1.25	Does not burn at all, only glimmers in a separate flame, burning zone covers with white deposit, white fumes	Highly flexible and elastic		No	Rubber-like to lardy	Flexible tubings, self-releasing molds, joint-filling materials	Whitish, gray, cream, beige, sometimes other colors (red, black, etc.)

The most important resin-based adhesives

The use of plastic resins for glueing has become a common practice for many amateurs today, who are quite unaware of this fact. This applies to the widely used white glue (based on polyvinyl acetate), to the large group of contact adhesives (based on synthetic rubber or polyurethane), as well as to the popular two-component adhesives, which are quite often based on epoxy resin but may sometimes also be based on vinyl resins, which form so-called copolymers or compounds with other types of resin.

In the first place, two-component adhesives are highly esteemed by both amateurs and professionals, because they allow them to bond even metal and other nonporous materials safely together and to produce durable joints of amazing strength. Besides these ready-to-use two-component glues, you may also use some liquid resins for glueing. Epoxy, PU and silicone rubber resins are especially suitable for this purpose. As with all types of glueing processes, a careful removal of any grease and, as far as possible, a good roughening of the bonding areas are indispensable. Epoxy resins are ideal adhesives for metal, ceramics, glass, stone, concrete, china—in short, for any nonporous and nonabsorbent materials. Silicone rubber is an excellent glue for glass, which even enables you to make an aquarium from five sufficiently thick glass panels without the need of any supporting metal frame. Just glue the glass panels together with silicone rubber. Silicone rubber also proves to be suitable for glueing acrylic glass to silicate glass (window glass), plastics, brickwork and wood, as well as metal to ceramics.

How to treat plastic furniture. Larger and larger quantities of plastic furniture are put on the market. The collective name "plastic" furniture comprises five different materials: glass fiber-reinforced polyester resin, PU integral foam, integral polystyrene foam, vacuum-formed or injection-molded polystyrene or ABS for small-size furniture and, finally and fairly rarely, parts from hot-molded acrylic resin.

Susceptible to scratches

All these different kinds of plastic furniture have two important things in common: their surfaces are susceptible to scratches and, therefore, cannot withstand abrasive cleaning agents. In addition, they tend to accumulate static electricity, which makes them attract dust and dirt.

This leads to two consequences, as far as cleaning is concerned. One should only use very mild cleaning agents, which Use only mild cleaners must by no means contain any abrasive substances. The most suitable cleaner is plain soap. This has a favorable additional effect: if you do not rinse the furniture with clear water but instead allow the slightly soapy water to dry and then polish the surface with a soft rag, the remnants of soap produce a faint invisible humid film on the surface of the plastic furniture; this renders the surface conductive and, thus, reduces the trouble of static electricity, which sometimes bothers sensitive people, as it may cause a harmless but unpleasant electric shock. The effect will, of course, not last forever but fades in the course of time, so that another treatment with soapy water is required after some time. There also are antistatic fluids and sprays that work on the same principle, i.e., making the surface conductive. If it happens that slight scratches impair the finish of plastic furniture, special polishing pastes that contain protective waxes can be used to remove them. This works very well with glass fiber and acrylic resin furniture but quite often leads to good results with other plastics, too. Make sure when buying the polishing wax that it is the right type, i.e., that it matches the hardness of the plastic material you want to touch up; otherwise, you may cause more damage to the finish than there was before.

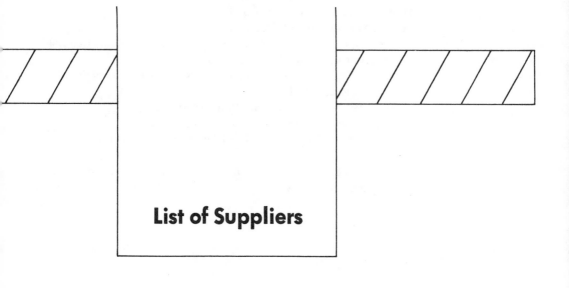

List of Suppliers

ACETATE FILM OR FOIL: model shops; plastics supply shops

ACRYLIC PANELS (PLEXIGLAS, PERSPEX): Rohm and Haas; Cadillac Plastic & Chemical Co.; William Cox Ltd.; Cornelius Chemical Co.; plastics supply shops

CAR BODY REPAIR KITS AND POLYESTER FILLER: garages; motor shops; hardware stores

DECORATIVE LAMINATES: wood dealers; hobby shops; hardware stores

EMBEDDING KITS, RESINS AND CRACK-EFFECT PASTE: plastics supply shops; hobby shops

EPOXY FLOORING MATERIALS: hardware and paint stores

GLASS FIBER MAT, FABRIC, STRANDS: Owens-Corning Fiberglas; PPG Industries; Bondaglass-Voss Ltd.

LAMINATING TOOLS: Automatic Process Control; Bondaglass-Voss Ltd., Strand Glass Co.; K & C Mouldings

MICROSPHERES: Microbeads Cataphote Division, Ferro Corp.; Emerson & Cuming; Fillite (Runcorn) Ltd.

MIXING AND PROPORTIONING DEVICES AND TOOLS: Ramco Industries; Bondaglass-Voss Ltd.

PEBBLES: garden equipment shops

PLASTIC LINING MATERIALS: wood dealers

PLASTIC VENEERS: wood dealers; hobby shops

POLYESTER, EPOXY AND GELCOAT RESINS: Diamond Alkali; Reichold Chemical; Bondaglass-Voss Ltd.

POLYSTYRENE, PVC AND PU FOAM BLOCKS AND PANELS: dealers for building materials

PU FLOORING RESIN: Dura Seal; Bondaglass-Voss Ltd.

THERMOPLASTIC FILM OR FOIL: plastics supply shops

THERMOPLASTIC GRANULATE: local injection-molding firms; hobby shops

WELDING MACHINES FOR PLASTIC FILM OR FOIL: hardware stores; department stores

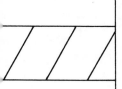

Trade Names
of Plastics and
What
They Actually Mean

(Abbreviations as used in table on pages 252–259)

Absinol – ABS
Abson – ABS
Acelon – CA
Acpol – UP
Acrylite – PMMA
Aerolite – UF
Alathon – PE
Alkathene – LDPE
Amberol – PF
Ampol – CA
Araldite – EP & PF
Aroset – PMMA
Arothane – UP
Arpylene – ABS

Bakelite – PF, EP, HDPE,
    LDPE & PF

Beckophen – PF
Beetle – PVC & UP
Beetle Resins – UF
BP – PS
Breon – PVC

Capron – PA
Carinex – PS
Carlona – HDPE & PP
Castomer – PU
Cellidor – CA
Cellobond – UP
Chemfluor – PTFE
Chem-o-sol – PVC
Chem-o-thane – PU
Cordo – PVC
Co-Rezyn – UP

Corvic – PVC
Crystic – UP
Cyanaprene – PU
Cycolac – ABS
Cymel – MF

Derakane – PVC
Dexcel – CA
Diakon – PMMA
Diaron – MF
Dion – UP
Drisil – SI
Durethene – PE
Dylan – PE
Dylene – PS
Dylite – EPS

Eanplast – CA
EAP – PS
Epikote – EP
Epi Rez – EP
Epocap – EP
Epok – PF
Epotuf – EP
Estane – PU
Ethafoam – PE
Evacor – PA
Extru-foam – PVC

Fassloid – PVC
Filabond – UP
Fluon – PTFE
Formica – MF
Formid – ABS & PP

Geon – PVC

Halon – PTFE
Hostaset – MF & UF

Impol – UP
Intamix – PVC
Isomin – MF
Isonate – PU

Jablite – EPS

Kollercast – EP
Kralastic – ABS
Kralon – ABS & PS
Krystal – PVC

Laminac – UP
Lexan – PC
Lucite – PMMA
Luran – ASA
Lustrac – CA
Lustran – ABS & ASA
Lustrex – PS

Makrolon – PC
Marlex – PE, PP & PS
Meldin – PA
Melmex – MF
Melolam – MF
Merlon – PC
Microthane – PU
Minlon – PA
Monsanto – HDPE & LDPE
Moplen – PP
Moranyl – PA
Multron – UP

Nestorite – UF
Nimbus – PU
Nypelube – PA

Olefo – PP
Oroglas – PMMA

Palatal – UP
Paraplex – UP
Paxon – PE
Pellethane – PU
Petrothene – PE
Plaskon – PA & UF
Plaslube – PC
Plastazote – PE
Plenco – MF & UF
Plexiglas – PMMA
Plyophen – PF

Polyfoam – PU
Polygrain – EPS
Polyite – UP
Polylube – PTFE
Pro-Fax – PP
Propathene – PP

Resimeme – MF
Resinox – PF
Resophene – PF
Rhodoid – CA
Rigidex – HDPE
Rilsan – PA

Silastic – SI
Sta-Form – UF
Sternite – ABS, EP, PF, PS & UF
Styrocell – EPS
Styrodur – PS
Styrofoam – PS
Styropor – PS
Super Dylan – PE
Swedcast – PMMA

Synolac – UP
Syn-U-Tex – MF & UF

Technyl – PA
Teflon – PTFE
Tenite – CA
Thurane – PU

Uformite – MF & UF
Ultramide – PA
Uvex – CA

Varcum – PF
Vibrathane – PU
Vibrin-Mat – UP
Vinaliner – PVC
Vinatex – PVC
Vydyne – PA

Wellamid – PA
Welvic – PVC

XT polymer – PMMA

Zelux – PC
Zytel – PA

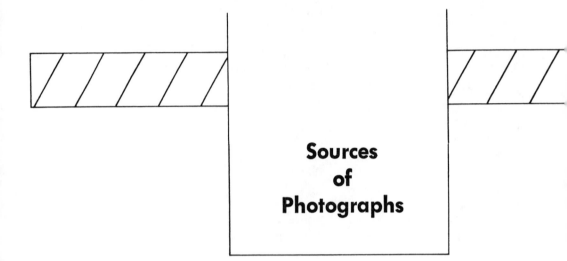

Sources
of
Photographs

BASF, Ludwigshafen (L. Peinecke, Frankenthal; Prof. Max Hunzicker, Zürich) (5), Bayer AG, Leverkusen (1), photographic library of Bertlesmann, Gütersloh (1); Bostik GmbH, Oberursel/Taunus (2), Dynamit Nobel AG, Troisdorf (1); Emerson & Cuming Inc., Canton (Mass.) (2); Gevetex-Textileglas-GmbH, Düsseldorf (1), Erich H. Heimann, Düsseldorf (22); Heinze (Lemgo) (1); Farbwerke Hoechst AG, Frankfort-Hoechst (3); Emil Lux, Wermelskirchen (1): G. Neuhaus, Frankfort (Main) (1), Omni-Technik GmbH, Munich (2); Vosschemie, Uetersen & Bondaglass-Voss Ltd., Beckenham (Kent) (17).

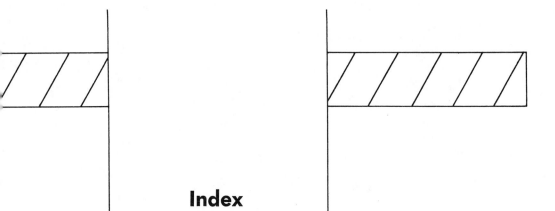

Index

accelerators (for hardening
process), 18–19, 25, 26,
121

acetone, 46, 58, 83, 106, 245

acrylic resins and glass, *see
also* Perspex; Plexidur;
Plexiglas, 209, 237
adhesives for, 245, 247–
49, 260
bending of, 212–13
characteristics of, 253
drilling of, 233
filing of, 235
for floors, 113
milling and engraving of,
235
polishing of, 236
sanding of, 235, 236
sawing of, 231–32
scraping of, 235
suppliers of, 263
uses of, 253
acrylonitrile-butadiene-
styrene (ABS):

adhesives for, 247
characteristics of, 252
for furniture, 260
uses of, 252
adhesives, 35, 242–51
bonding key, 247–50
contact, 239, 240, 243–44,
246, 247–49, 260
dangers of one-component
reactive, 245
emulsion-type, 244, 247–
49
for ironing decorative laminated panels onto wood,
244
list of available, 246–47
for mending of textiles
with patches, 244
permanently tacky, 245–
46
reactive, 245, 247–49
resorcinol, 246, 247–49
stresses on, 243
swelling, containing sol-
vent, 244, 247–49

two-component reactive,
246, 247–49, 251, 260
white glue, 45, 46, 163,
193, 227, 247, 260
aerosil paste, 111
aerosil powder, 111, 120
aircraft, *see also* fuselages,
32–33
model, 161–62, 226–27
amber resin, 120–21
aquarium, making an, 260

Baekeland, Leo Hendrik, 13
Bakelite, 13
Bayer, Otto, 125–26
Bayer Company, 125–26,
145
bending, 208, 212–15
of acrylic glass, 212–13
forming temperature and,
215
properties of thermoplas-
tics and, 214–15
benzene, 217, 218, 248
Berite, 237

polycarbonates (*cont.*)
adhesives for, 248
characteristics of, 253
filing of, 235
polishing of, 236
sanding of, 236
uses of, 253
polyester filler, 47, 66, 67, 79, 85, 86, 163, 228
components of, 97
mixing of, 97
for models, 100
pastes, 98
pot life of, 98
properties of, 197–98
for repairing car bodies, 99, 263
as sealant, 100
spray variety, 98
tinting, 99
polyester resins, *see also* unsaturated polyester resins, 13, 15, 18, 55, 61, 124, 144, 168
furniture from, 4, 13, 26
hardener and, 29–30
for potting electronic circuits, 122
as primers, 83
properties of, 26–29, 79
for reactive adhesives, 245
sealing coat needed with, 28–29, 43, 89, 115
shrinkage of, 26–27, 108
ultraviolet-sensitized, 26
polyester resins and glass fiber, 26, 32–34, 35, 163
for boats, 65, 68, 69, 70, 228
for coating of surfaces, 78–80, 158
for coating terraces, 84–86
as corrosion reducer for steel boats, 84
for furniture, 74–75, 260
for lamps, 75–76
for mending rusted car bodies, 80–81
for models, 74
for molds, 48
for repairing leaky flat roofs, 80, 86–87
for sealing cracked concrete, 80
for sealing old wooden boats, 80, 82–84
for swimming pools, 80, 86–95, 205
wall reliefs made from, 77
polyester resins and metal or

ceramic powder fillers 100–106
choosing the filler material, 103–104
compared with epoxy resin and like fillers, 102
preparation of, 103–104
properties of, 101–103
uses of, 106
polyester resins and microspheres, 107–108
advantages of, 106
for imitation wooden panels, 108
suppliers of, 263, 264
polyester resins and stone or pebbles, 108–11
to simulate stone or marble, 108
suppliers of, 264
uses for, 109–11
polyester resins and wood filler, 106–107, 170, 171
polyester resins for embedding process, 114–21
casting process for, 117–18
finishing touches for, 118–19
molds for, 116–17
objects to be made from, 121
use of crack-effect paste with, 121
polyethylene, *see also* polythene:
adhesives for, 248
flexible, characteristics of, 253
rigid, characteristics of, 253
scraping of, 236
uses of, 6, 252, 253
polyhydric alcohols, 126
polymerization, 21, 25
polymethyl methacrylate, *see* acrylic resins and glass
polyolefin plastics, *see also* polypropylene; polythene, 240, 241
glueing of, 246
polypropylene:
adhesives for, 248
bowls, 4
characteristics of, 254
cutting of, 232
impossibility of painting, 240
molds, 116
planing of, 235
uses of, 7, 251
polystyrene:
adhesives for, 248

bonding of, 245
characteristics of, 254
cutting of, 232
filing of, 235
for furniture, 260
impressed, film or foil for veneers, 194
for "melted pictures," 217
for models, 241
painting of, 240–41
sanding of, 235
temperature and, 10, 208, 215, 216
uses of, 6, 254
for veneers, 193
polystyrene cement, 223, 224
polystyrene foam, 11–12, 45–46, 77, 144, 163, 191, 219, 248
adhesives for, 45, 248
cutting of, with electrically heated steel wires, 219, 222, 232
epoxy resins and, 46
extruded, 219–20, 221
fire resistant, 219, 221
for furniture, 12, 223, 260
for insulation, 11, 220
for modeling, 225–27
for packing, 11
painting of, 240–41
panels, 219, 220, 222, 224, 241
sanding of, 236
shelves, 223
polysulfides, 137, 138
polytetrafluoroethylene:
adhesives for, 249
characteristics of, 255
polythene, *see also* polyethylene:
adhesives for, 246, 248
bags, 143, 152
impossibility of painting, 240
lamps, 225
"melted pictures" from, 217
panels, cutting of, 232
planing of, 235
repairing, with permanently tacky adhesives, 246
in swimming pools, 93
temperature and, 10, 216
tubing, 137
uses of, 6, 19, 116
welding of, 242
polyurethane (PU) compounds, 77, 127, 139
advantages of, 139
for filling of joints, 14, 135–39

thermoplastics (*cont.*)
properties of, 6, 7–8, 213
PVC tiles for floors, 195–97
rasping of, 235
sanding of, 235–36
sawing of, 231–32
scraping of, 235
self-adhesive sheetings of, 190–92
sheeting, use of, 197, 202
sheeting, welding of, 198–204
softening point of, 208
sources of heat for forming, 209–10
tapping of, 237
tubings, 190
turning on a lathe, 233
uses of, 5, 6, 11
veneers, 193–94
welding of, 210–11
thickening agents, 111
tools, *see names of specific tools*
trichlorethylene, 217
trichloro-ethane, 55
tri-ethane chloride:
for cleaning tools, 58
two-component epoxy resins:
as sealants, 171
two-component liquid plas-

tic resins, 15, 17–23, 143–44
mixing proportions, 17–19, 23
purchasing, 23
safety precautions for working with, 22, 23
shrinkage of, 21
stirring and mixing of, 19–22, 23
storage of, 23
two-component polyurethane (PU):
for casting massive casts and molds, 145
for coating of floors, 140–42
for coatings, 87, 140–43
for coatings of flat roofs, 142–43
effects of temperature on, 140–41
for filling joints, 143–44
for glueing, 144
primer for, coatings, 134
properties of, 140
safety precautions for working with, 140
uses of, 139–40
two-component PU lacquers, 145–46
unsaturated polyester resins, 15, 24–29

accelerators for, 25
adhesives for, 249
characteristics of, 258
curing temperature, 25
hardening systems, 25
mixing, 25–26
safety precautions for working with, 24–25
shrinkage of, 21, 26
uses of, 24, 258
urea formaldehyde, 170, 171, 172
adhesives for, 249
characteristics of, 257
uses of, 257

vinyl acryl resins:
for reactive adhesives, 245

welding, of thermoplastic sheetings, 198–204, 242
chemical, 203–204
with an electric iron, 198–201
with special welding equipment, 201–203, 264
temperature for, 202
"white fracture," 194, 208, 211
white glue, 45, 46, 163, 193, 227, 247, 260